T5-AMY-968

No. 2625
$23.95

HOW TO USE
SPECIAL-PURPOSE
ICs

DELTON T. HORN

TAB BOOKS Inc.
Blue Ridge Summit, PA 17214

FIRST EDITION

FIRST PRINTING

Copyright © 1986 by TAB BOOKS Inc.

Printed in the United States of America

Library of Congress Cataloging in Publication Data

Horn, Delton T.
    How to use special-purpose ICs.

    Includes index.
    1. Integrated circuits. I. Title.
TK7874.H676   1986   621.381′73   86-6006
ISBN 0-8306-1325-0
ISBN 0-8306-1725-6 (pbk.)

Contents

Introduction

I NTEGRATED CIRCUITS ARE A VITALLY IMPORTANT PART OF MODERN electronics. No one who works with electronics, either professionally or as a hobby, can afford to ignore these incredible devices. The miniaturization of ICs makes countless applications practical and inexpensive that could not have been seriously considered just a few years ago. Who could have afforded a home computer in 1965? Or even have the room to store it?

Keeping up with ICs is difficult. There are thousands of IC devices now on the market, and dozens of new chips are being announced every month. Many are primarily improvements on older devices. This is especially true for general-purpose ICs, like op amps.

There is plenty of literature available on general purpose ICs. For example, two of my own books, *How To Design Op Amp Circuits, With Projects and Experiments* (TAB book #1765), and *Using Integrated Circuit Logic Devices* (TAB book #1645) fall into this category.

Many ICs are designed to perform specific functions. Because of their more limited applications, less has been written about them. This book is intended to give the reader at least a passing familiarity with most of the broad areas of functions that can be performed by special-purpose ICs. This is a huge territory, ranging from simple voltage regulators to complex CPUs.

There is obviously no way for this book to be completely comprehensive. For one thing, new products are being released even as this book is coming off the presses. I have tried to give a reasonably comprehensive overview of a number of general areas.

Another problem with a book of this nature is one of organization. Traditionally a distinction has been made between linear (analog) and digital circuitry. With many special-purpose ICs the line blurs. Many chips use digital circuitry to perform analog functions. I have tried to use the intended application as the deciding factor. The first twelve chapters deal with linear functions, and the last six chapters cover devices intended for digital applications. Bear in mind that there is considerable overlap between these two areas. In many cases the choice of where to discuss a specific IC became rather arbitrary.

I have made every effort to make this book more than just another collection of data sheets. Whenever possible, I have tried to give some practical information on the use of many of the chips discussed. It isn't practical to go into depth on every device. The book would end up running into several volumes.

Let me repeat, this is an overview—not the final word on the subject. I sincerely hope it will make it easier for you to work with special-purpose ICs and apply them in your own specific applications.

```
    ┌───┬─U─┬───┐
    │ 1 │ S │24 │
    │ 2 │ P │23 │
    │ 3 │ F │22 │
    │ 4 │ C │21 │
    │ 5 │ I │20 │
    │ 6 │ A │19 │
    │ 7 │ L │18 │
    │ 8 │ P │17 │
    │ 9 │ U │16 │
    │10 │ R │15 │
    │11 │ P │14 │
    │12 │ O │13 │
    │   │ S │   │
    │   │ E │   │
    └───┴───┴───┘
```

Chapter 1

Power Supply ICs

C ERTAINLY THE MOST COMMON OF ELECTRONIC APPLICATIONS IS THE power supply. All circuits require some source of power. Even passive circuits, such as simple RC filters, steal their operating power from the signal passing through them. A great many special-purpose ICs have been developed for the prosaic, but vital power-supply functions.

VOLTAGE REGULATORS

A voltage regulator is a circuit that smooths out a voltage to a precise, non-varying dc voltage. It removes any noise or ac component (ripple). It also prevents fluctuation of the supply voltage with changes in the load current.

Voltage regulator ICs are available for a wide variety of voltages, and current-handling capabilities. They may be designed for either positive or negative voltages (with respect to ground). A single unit usually cannot be employed for both positive and negative operation.

These devices are certainly easy to work with, at least in basic circuits. Generally, they have only three leads. In fact, they often resemble a slightly over-sized power transistor. The three leads of a typical voltage regulator IC are usually identified as follows:

- ☐ Unregulated input voltage
- ☐ Regulated output voltage
- ☐ Common

The common connection, of course is the nominally ground reference point for both the input and output voltages.

1

Some typical voltage regulator ICs are illustrated in Fig. 1-1. Figure 1-2 shows the basic circuit for using a typical voltage regulator IC. The four capacitors perform various filtering functions. Capacitor C1 is a large, brute force filter with a high capacitance value. Its purpose is to smooth out the ripple from the rectifier output. You could think of it as sort of a "pre-regulator" to partially condition the voltage before it reaches the actual regulator IC.

Capacitors C2 and C3 have relatively small values—usually under 1 microfarad. These capacitors help the regulator handle local transients, especially if the regulator is physically located some distance from C1 and C4. Capacitors C2 and C3 are placed as close to the regulator IC as possible. Often they will be mounted directly on the leads of the IC.

Finally, C4 takes care of sudden surges that may show up on the V+ output line. These surges can show up especially when the system is powered up or powered down. The size of the surges will depend on the current draw of the system. This will determine the ideal size of C4. Typically, this capacitor will have a value in the 10 to 100 μF range.

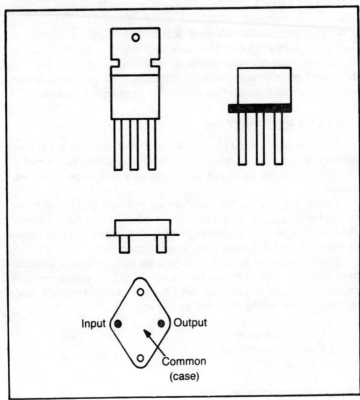

Fig. 1-1. Voltage regulator ICs come in several package sizes.

2

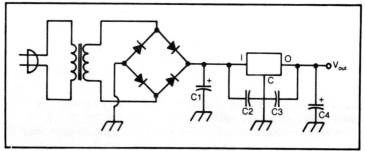

Fig. 1-2. Voltage regulators are easy to work with.

The LM309

One of the most popular and widely available voltage regulator ICs is the LM309. This is a standard three-pin device, designed for a +5 volt output. The unregulated input voltage may go as high as +35 volts.

The LM309 can handle a fairly hefty amount of power. Even without any external heatsinking, it can dissipate more than 3 watts at room temperature without running into trouble. By adding a good heatsink, like the Wakefield 680-75, the power handling capability of this device can be raised to over 10 full watts.

The user doesn't even have to be too concerned about these power limits, since the chip includes internal protection. An automatic sensor circuit within the IC will shut down the unit if the chip temperature goes above a safe limit. Excess power dissipation will heat up the IC, of course, and this will cause the regulator to shut itself down before damage occurs. This IC is available in several package types, with standard TO-3 and TO-39 cases being the most common.

Base diagrams for the LM309 are shown in Fig. 1-3. The transformer secondary is rated for 6.3 volts, or higher (up to about 30 volts). A diode bridge rectifier converts the ac voltage into a pulsating dc voltage. Either separate diodes, or a modular bridge rectifier unit may be used here. The

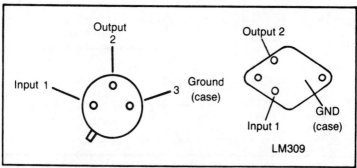

Fig. 1-3. The LM309 is a popular +5-volt regulator.

PIV (Peak-Inverse-Voltage rating) should be in the 50- to 200-volt range. The capacitors, of course, serve as filters.

The output of this circuit is a well-regulated voltage with a nominal value of + 5.05 volts. There is a ± 0.2 volt tolerance, so the actual output voltage may range from + 4.85 volts to + 5.25 volts. This should be accurate enough for the vast majority of applications you are likely to encounter. This circuit is ideal for use with TTL devices, which are notoriously fussy about their supply voltages.

A simple resistive voltage divider across the output, as shown in Fig. 1-4, allows you to vary the output voltage across a 0- to 5-volt range, while maintaining good regulation.

The 78xx Series

A popular series of IC voltage regulators is the 78xx series. These three-pin devices, which are illustrated in Fig. 1-5, are available in a wide variety of frequently used voltages. The voltage is identified in the last two digits of the type number. For example, the 7805 is a + 5-volt regulator, and the 7812 is a 12-volt regulator.

All of the 78xx series voltage regulators are designed for positive voltages. Negative voltage equivalents are offered in the 79xx series. The 7905, for example, is the same as the 7805, except it puts out – 5-volts regulated, instead of + 5 volts. The 78xx (and 79xx) series of regulators can typically dissipate about one ampere.

Increasing Current-Handling Capability

While modern ICs are being designed to draw less and less current,

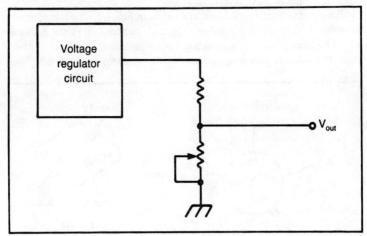

Fig. 1-4. A simple resistive voltage divider across the output of a voltage regulator can reduce the output voltage.

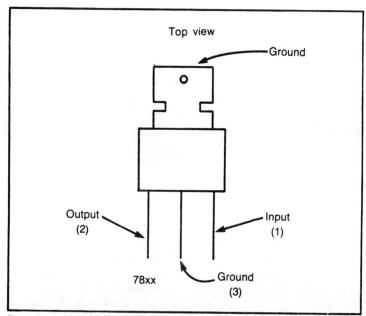

Fig. 1-5. The 78xx series of voltage regulators are available in a number of popular voltages.

the current draw in a complex circuit with many ICs can quickly add up to a significant amount. In addition, many off-chip devices (such as LEDs, and relays) are notorious power hogs. Also, when heavy amounts of current run through the copper traces of a PC board, the resistance of the trace itself can start affecting circuit operation. Parasitic oscillations are one frequent problem that can be caused by trace resistances.

The result of all this is that often the one amp provided by a 78xx series regulator just isn't enough to do the job. Fortunately, there are ways to get around the internal current limitations of the voltage regulator IC. We will be using the 7805 in our examples, but the basic principles can be applied to virtually any voltage regulator IC.

One way to improve the current-handling capability of a voltage regulator IC is to add a pass transistor and resistor, as shown in Fig. 1-6. The current flowing through the regulator must come from the supply voltage ($+V_{in}$), of course. In this circuit, before the incoming current can get to the regulator, it must first pass through the resistor (R_b). As the current flowing through this resistor increases, so does the voltage drop across it. This is in accordance with Ohm's law ($E = IR$).

The base/emitter junction of the pass transistor is in parallel with resistor R_b. Ordinarily, the voltage drop across this resistor will be too low to turn the transistor on, so it just sits there, as if it wasn't in the circuit at all. The voltage regulator functions normally. At some point, however, the

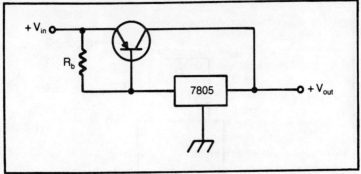

Fig. 1-6. Adding a pass transistor can increase the current handling capabilities of a voltage regulator.

voltage drop across the resistor will be large enough to turn the transistor on. Now the transistor will start passing current through its collector, which is tied to the output of the voltage regulator IC. The transistor current is added to the regulator current. In effect, the current-handling capability of the voltage regulator is increased by an amount equal to the current-handling capability of the pass transistor. Since power transistors that can handle several amps are available, this can mean a significant increase in current-handling capability with the addition of just two components.

There is a trade-off involved though—it seems there always is. The difference between the input and the output voltage will change. As you know, the input voltage supplied to the regulator must be somewhat higher than the desired output voltage. The transistor and resistor add about a 2-volt drop. If the straight regulator circuit required, say, a 6-volt input for a 5-volt output, the addition of the pass transistor and resistor would require an 8-volt input for the same 5-volt output. This is rarely a major problem, but you should be aware of it.

A more serious problem is that if the output is shorted, or starts drawing an excessive amount of current for any reason, the pass transistor will be quickly destroyed. Sure, the voltage regulator IC generally has short-circuit protection, but it only protects the IC itself. Any external components are on their own.

A solution to this problem is illustrated in Fig. 1-7. Here we've added another transistor/resistor combination. Notice that the new transistor (Q2) and resistor (R_s) are in the same configuration as the original pass transistor (Q1) and resistor (R_b). They act as another current-sensing switch.

Incoming current has to flow through R_s before it can get to Q1. Ordinarily, the voltage drop across R_s won't be enough to turn Q2 on, so this second transistor is effectively out of the circuit. If Q1 tries to draw too much current, the voltage drop across R_s will become sufficient to turn on Q2. This will forcibly lower the voltage drop across R_b, and turn Q1 off before damage can be done. Because Q2, by definition, will only be

6

on when there are relatively large currents flowing, it should be a very heavy-duty power transistor. After all, if Q2 burns out, there goes our short-circuit protection.

Now, how do we go about choosing the values for R_b and R_s? For simplicity we will ignore Q2 and R_s for the time being, and just work with the circuit shown in Fig. 1-6. The first step is to determine the turn-on voltage for the transistor. For silicon transistors (the most common type today) this will be about 0.7 volt.

Next, we need to decide how much current flow should turn on the transistor. By itself, a 7805 voltage regulator IC needs about 8 mA (0.008 amp) to operate. All other current will be passed on to the load circuit. A 78xx series device is rated for up to 1 ampere. Let's leave ourselves a wide margin for error, and say we don't want more than 0.5 amp (500 mA) to flow through the regulator chip. This way we can be sure that the IC won't be overstressed. Now, we add the operating current (0.008 amp) to the maximum flow current (0.500 mA). We want the voltage drop across R_b to be enough to turn on Q1 when the current flow exceeds 508 mA (0.508 amp). In other words, when the current through R_b is 0.508 amp, the voltage drop across it should be 0.7 volt. We can use Ohm's law to find the desired resistance:

$$R = E/I$$
$$R = 0.7/0.508 = 1.4 \text{ ohms}$$

We can round this off to 1.5 ohms. The value calculated in this manner is not strictly accurate, because it ignores the base/emitter resistance of Q1, which is in parallel with R_b. Fortunately, the base/emitter resistance

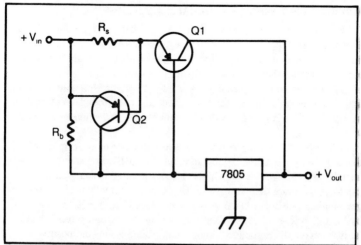

Fig. 1-7. A second pass transistor can be used to add short circuit protection.

when the transistor is cut off is quite high, so it will have very little effect compared with the very low resistance of R_b. We can ignore it with reasonable confidence that it won't make any noticeable difference.

In the short circuit protected circuit of Fig. 1-7, things get a bit more complicated, because R_b, R_s, and Q2 are all in parallel with the base/emitter junction of Q1, so they all have to be considered. Q2 can be safely ignored here, since its value will be so large with respect to the two resistors. We only need to concern ourselves with R_b and R_s.

The value of R_s can be found with the straightforward application of Ohm's law presented earlier. First we need to decide on the maximum output current before Q2 will turn on, cutting off Q1. Let's say we want a maximum current-handling capability of 4 amps. This means the resistance of R_s should be approximately.

$$R = E/I$$
$$R = 0.7/4 = 0.175 \text{ ohm} = 0.18 \text{ ohm}$$

Now, we just subtract this from the calculated value for R_b to find the compensated value for this resistor.

$$R_b = 1.378 - 0.175 = 1.203 \text{ ohm}$$

This can be rounded off to 1.2 ohms.

Notice that very small resistances are required in this application. Some oddball values will often result in the calculations. Always double-check to make sure that rounding off the values won't result in the current-handling capabilities of any of the components being exceeded.

Making Fixed Voltage Regulators Variable

Fixed voltage regulator ICs are readily available in a number of standard voltages. Unfortunately, real-world circuits often demand nonstandard voltages. You can trick a fixed voltage regulator IC into putting out something other than its rated voltage.

It's easy enough to create a lower output voltage with a simple voltage divider network across the V+ output line and ground, as illustrated in Fig. 1-4. The ratio of the two resistor values determines the output voltage.

It's important to bear in mind the fact that most voltage regulator ICs, like the 78xx series units, require an input voltage that is somewhat higher (usually at least about 2 volts higher) than the rated output voltage. You can't get out more than you put in. Or can you? Actually it is possible to get a higher than normal voltage out of a voltage regulator IC. This is done by "lying" to the regulator about just where the ground reference point is. If we make the regulator "think" that ground is at some positive voltage above the true ground point, the rated output voltage will be added to the artificial ground point.

For instance, let's say we are using a 7805 voltage regulator IC, which is designed to put out $+5$ volts, and we connect $+1$ volt to the ground terminal of the IC. The output will be $+6$ volts. One way of implementing this technique is shown in Fig. 1-8. The two resistors (R1 and R2) form a voltage-divider network, defining the artificial ground. The output voltage for this circuit can be determined with this formula:

$$V_o = V_r + (V_r R2) / [(R - 1) + I_{sb}]$$

where V_o is the resulting output voltage, V_r is the rated output voltage of the regulator, and I_{sb} is the standby current used by the regulator.

For the 7805, V_r is $+5$ volts, of course, and I_{sb} is about 8 mA (0.008 amp) when there is no load. If we assign a value of 1 kΩ (1000 ohms) to R1, and 390 ohms to R2, the output voltage with no load would be:

$$
\begin{aligned}
V_o &= 5 + (5 \times 390) / (1000 - 1) + 0.008) \\
&= 5 + 1950 / (999 + 0.008) \\
&= 5 + 1950 / 999.008 \\
&= 5 + 1.95 \\
&= 6.95 \text{ volts}
\end{aligned}
$$

This approach works fairly well for applications with relatively low currents. But I_{sb} will change as the current draw increases. This will affect the output voltage, resulting in poorer regulation.

Another source of instability in this circuit is the power handling capability of the regulator. As the maximum current value is approached, the

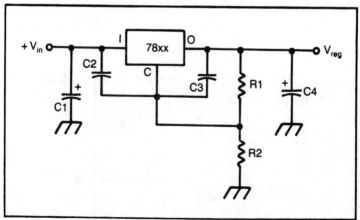

Fig. 1-8. A voltage regulator IC can be "tricked" into putting out a voltage higher than its nominal value.

resistors will start to get warm, causing their values to change—which changes the output voltage. At least the internal protection of the IC will prevent the thermal runaway that could develop in a transistor circuit, but the regulator will tend to drop-out prematurely, requiring you to overdesign it for greater current handling than is really needed.

A better approach to getting a higher than normal voltage out of a fixed voltage regulator is shown in Fig. 1-9. Here an op-amp buffer stage is added to prevent loading effects. The op amp is connected as a noninverting unity-gain voltage follower. The basic operating principles of this circuit are pretty much the same as in Fig. 1-8, but the high input impedance of the operational amplifier will improve the stability of the output voltage. (For more information on op amps, see my earlier book, *How To Design Op-Amp Circuits With Projects & Experiments*—TAB Book #1765).

The same tricks can be used to create lower than normal output voltages. Just apply a negative (below true ground) voltage to the common pin of the voltage regulator IC.

The TL431 Precision Adjustable Shunt Regulator

Most voltage regulator ICs are designed for a single output voltage. A few are designed to allow the user to select the output voltage. One such device is the TL431 precision adjustable shunt regulator. This chip is available in two standard package types—a three-lead TO-92 (see Fig. 1-10) and an 8-pin DIP (see Fig. 1-11).

The output voltage from the TL431 is programmable over a 2.5- to 36-volt range. Of course, the input voltage must be somewhat higher than the desired output voltage. This device can handle currents up to about 100 mA, and has a low dynamic-impedance (typically 0.22 ohms).

Besides basic shunt regulator circuits, such as the one shown in Fig. 1-12, the TL431 can be used in many related applications. For instance,

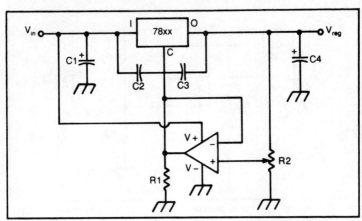

Fig. 1-9. Adding an op-amp buffer stage helps reduce loading effects.

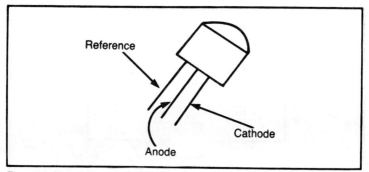

Fig. 1-10. The TL431 Precision Adjustable Shunt Regulator is available in a three-lead TO-92 can.

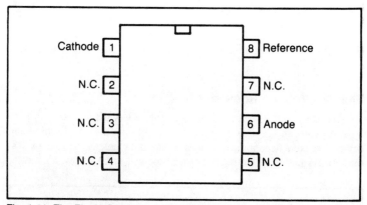

Fig. 1-11. The TL431 Precision Adjustable Shunt Regulator is also available in an 8-pin DIP housing.

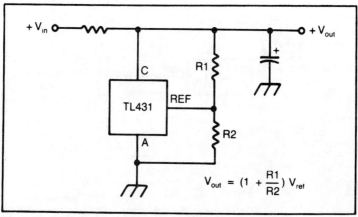

$$V_{out} = (1 + \frac{R1}{R2})\, V_{ref}$$

Fig. 1-12. This circuit is a basic shunt regulator using a TL431.

11

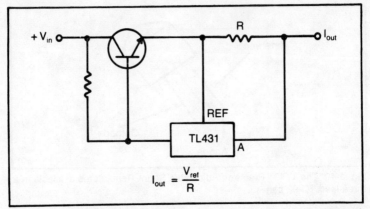

Fig. 1-13. The TL431 can also be used in such applications as a constant-current source.

Fig.1-13 shows a constant-current source, and Fig.1-14 illustrates a triac crowbar circuit.

The 723 Voltage Regulator IC

Another popular voltage regulator is the 723, which is shown in Fig. 1-15. As you can see, it is a bit more complex than the simple three-pin devices we have been looking at so far. One of the unique features of this device is its method of protection against excessive current flow. Most volt-

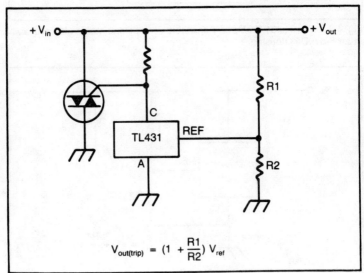

Fig. 1-14. This is a triac crowbar circuit built around the TL431.

12

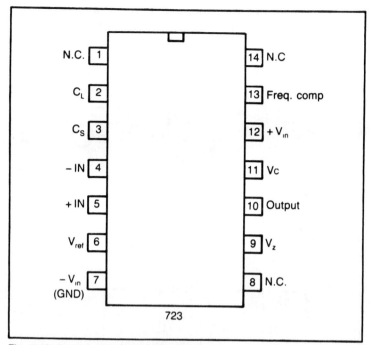

Fig. 1-15. Another popular voltage regulator IC is the 723.

age regulator ICs simply shutdown when too large a current demand is made upon them. The 723, however, utilizes a method called *current foldback*. This means when the maximum level is exceeded (due to a short in the load, or whatever), the 723 automatically drops the current output down to a small fraction of its output level. It simply refuses to put out any more. A functional block diagram of the internal circuitry of the 723 is shown in Fig. 1-16.

VOLTAGE CONVERTER ICs

In some applications it may be necessary to convert a positive voltage into an equivalent negative voltage. This can be done with a type of circuit known as a voltage converter. The ICL7660, shown in Fig. 1-17, is a voltage converter in IC form. A block diagram of this device's internal circuitry is shown in Fig. 1-18. The input voltage may range from +1.5 to +10.0 volts, resulting in a complementary output voltage (from -1.5 to -10.0 volts).

The ICL7660 is quite easy to use. Only a pair of external capacitors are required to make it operate (one from pin 2 to pin 4, and the other from pin 5 to ground). The capacitance values aren't even particularly critical. A basic ICL7660 voltage conversion circuit is illustrated in Fig. 1-19.

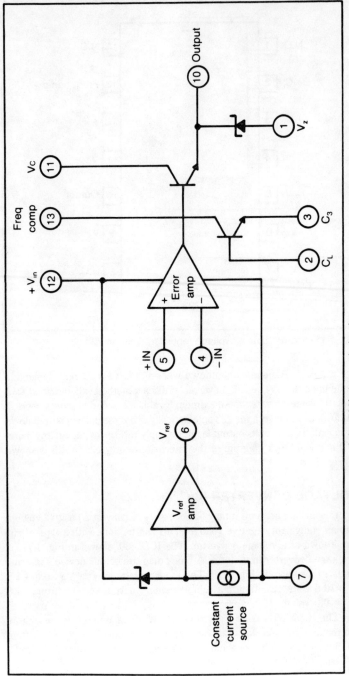

Fig. 1-16. This is a functional block diagram of the 723's internal circuitry.

14

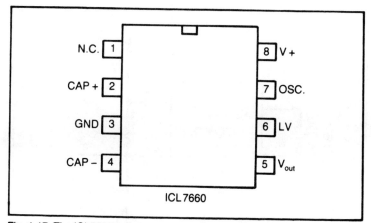

Fig. 1-17. The ICL7660 voltage converter IC is used to convert a positive voltage into an equivalent negative voltage.

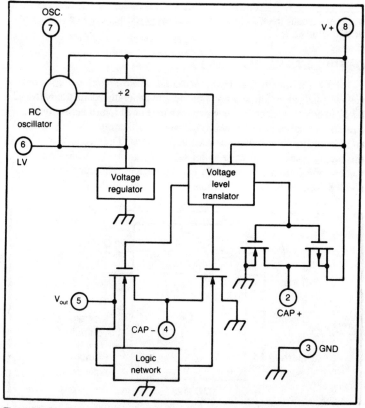

Fig. 1-18. This is a functional block diagram of the ICL7660's internal circuitry.

15

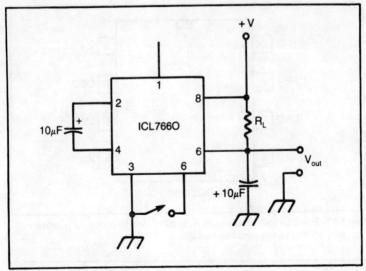

Fig. 1-19. Usually the ICL7660 voltage converter will be used in a circuit similar to this one.

OVERVOLTAGE SENSOR IC

Many ICs contain circuitry for protection against overheating, or overvoltage conditions. These protection circuits generally protect only the IC itself. In some applications, however, we may need to protect some sensitive off-chip circuitry. The MC3423, which is illustrated in Fig. 1-20, is an IC specifically designed for such applications. The MC3423 monitors the supply voltage, and triggers an external "crowbar" SCR circuit if a voltage transient or loss of regulation is detected, protecting any components driven by the supply voltage.

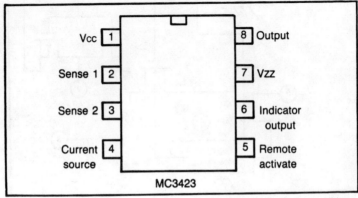

Fig. 1-20. The MC3423 protects external circuitry from overvoltage conditions.

PRECISION VOLTAGE SOURCES

In many electronics applications a very precise voltage is required as a reference to compare to an unknown signal. Such applications include calibrators for DVMs and oscilloscopes, A/D and D/A converters for computers, electronic thermometers, precision current sources, panel meters, and just about anything using a *Wheatstone bridge* (and that covers a lot of territory). The voltage regulators described earlier in this chapter are precise enough for power supplies, but they're not quite precise enough to be used as a high quality reference source.

Special precision voltage source ICs are available. A fairly typical example of this type of device is the LM199. It is offered in a round, 4-pin package, as shown in Fig. 1-21. The internal structure of this IC is illustrated in Fig. 1-22. Ignoring the two power supply pins, this is a two terminal device, that can be used in place of a zener diode. Basically it consists of a zener diode and an on-chip heating element (resistor).

The heater element in this IC is not actually a resistance element. It is a class A amplifier with the input shorted. This results in a more precise chip temperature, because the amplifier will dissipate a constant amount of power in this application.

An ordinary discrete zener diode is temperature sensitive, as are most electronic components. A drift specification of 5 mV per-degree-centigrade is typical. In the LM199 the zener diode's temperature is primarily determined by the internal heater element. Because they are on the same silicon chip, the heater and the diode must be the same temperature. The external ambient temperature is relatively insignificant. Besides improving the temperature stability, the on-chip heater also results in lower noise operation.

The zener diode in the LM199 is designed for a precise 6.95 volts. Pre-

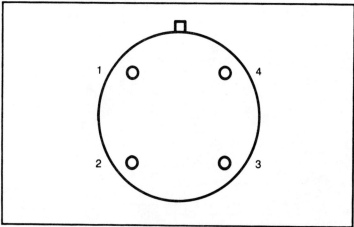

Fig. 1-21. The LM199 is designed to be a precise 6.95-volt reference.

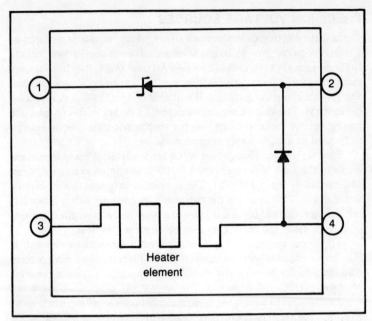

Fig. 1-22. The LM199 primarily consists of a zener diode with an on-chip heating element.

cision voltage source ICs for other voltages are also available. Precision Monolithics Inc. manufactures the REF-01 and REF-02 monolithic voltage reference ICs. The REF-02 puts out a stable +5.000 volts, and the REF-01 produces a precise +10.000 volts. The REF-02 also provides an additional output pin that puts out a voltage that is linearly proportional to the ambient temperature. This feature is not included on the REF-01.

Chapter 2

Phase-Locked Loops

```
     ┌───┬─┬───┐
    ╡ 1       24 ╞
    ╡ 2   S   23 ╞
    ╡ 3   P   22 ╞
    ╡ 4   E   21 ╞
    ╡ 5   C   20 ╞
    ╡ 6   I   19 ╞
    ╡ 7   A   18 ╞
    ╡ 8   L   17 ╞
    ╡ 9   P   16 ╞
    ╡10   U   15 ╞
    ╡11   R   14 ╞
    ╡12   P   13 ╞
         O
         S
         E
```

AN IMPORTANT TYPE OF CIRCUIT USED IN MANY APPLICATIONS IS THE *phase-locked loop*, or PLL. Unfortunately, many people who work with electronics fear the PLL as something mysterious and hard to understand. This even includes a number of experienced technicians. The PLL doesn't really deserve its arcane reputation. True, it is somewhat more complex than a basic oscillator or power supply, but it isn't all that difficult to understand, especially if you break it up into subcircuits. Working with PLLs is made even easier by their availability in IC form.

More and more commercial equipment is including PLL stages. There really isn't any excuse for anyone working with electronics in any capacity (professionally or as a hobby) not to have at least a nodding familiarity with the phase-locked loop.

WHAT IS A PHASE-LOCKED LOOP?

A basic phase-locked loop circuit is made up of three stages, as illustrated in Fig. 2-1. It is essentially a pseudo-servo-system with a controlling feedback loop. The output frequency locks onto and follows (or tracks) an input reference signal. The output is an integral multiple of the input reference frequency. Moreover, the output signal is in-phase with the input signal.

The VCO (*voltage-controlled oscillator*) generates a signal with some specific frequency. It may be higher than it should be, lower than it should be, or exactly on frequency. The VCO's output signal is fed back to the phase-detector stage, where it is compared with the input reference signal.

If the VCO is putting out the desired signal, it will be in-phase with the input signal, and the phase detector won't put out an error voltage.

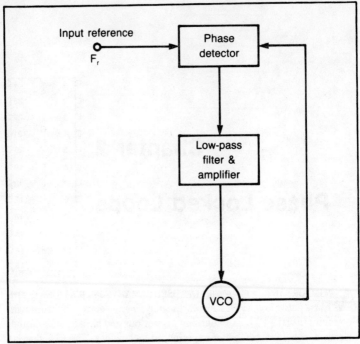

Fig. 2-1. A phase-locked loop is made up of three basic stages.

The VCO continues generating the output signal at the same frequency.

On the other hand, if the output from the VCO is at too low a frequency, it will be out-of-phase with the input reference signal. The phase detector will put out an error signal, which will be conditioned by the filter/amplifier stage. The conditioned error voltage will be fed to the control input of the VCO forcing it to raise the frequency of its output signal.

The same thing happens if the output frequency is too high, except the polarity of the error voltage is reversed, causing the VCO to decrease its input frequency. That's all there is to the basic PLL. It really isn't that complicated after all.

THE LM565 PLL IC

A typical PLL in IC form is the LM565. A block diagram of this device is shown in Fig. 2-2. It is available in several package types. Figure 2-3 shows the pinout diagram for the 14-pin DIP version, and Fig. 2-4 illustrates the round 10-pin package.

The LM565 is a general-purpose device, and can be used in a variety of applications. Some typical applications include:

☐ Data synchronization

20

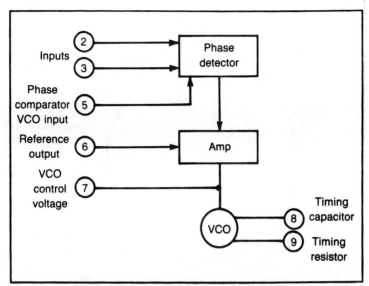

Fig. 2-2. A popular PLL IC is the LM565.

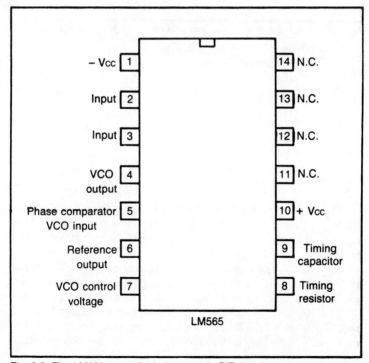

Fig. 2-3. The LM565 is available in a 14-pin DIP housing.

21

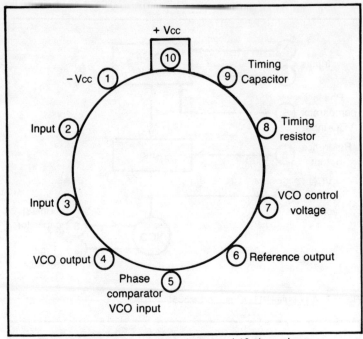

Fig. 2-4. The LM565 is also available in a round 10-pin package.

☐ FM demodulation
☐ Frequency division and multiplication
☐ FSK demodulation
☐ Modems
☐ SCA demodulators
☐ Signal regeneration
☐ Tape synchronization
☐ Telemetry receivers
☐ Tone decoding

The 565 is very flexible about its power supply requirements. It will run on anything from ± 5 volts to ± 12 volts, although a stable regulated supply is advisable for best results. The exact voltage isn't too critical, but it should be consistent.

The LM565 contains a stable VCO, which is designed for high linearity. Its basic frequency is determined with an external resistor and capacitor. A 10:1 frequency range can be achieved with a single tuning capacitor.

This chip also contains a double-balanced phase detector. The feedback loop between the VCO and the phase detector may be externally broken to add a digital frequency divider. This will allow the PLL to function as a frequency multiplier.

22

The characteristics of the closed loop system may be adjusted over a wide range with an external resistor/capacitor combination. The adjustable characteristics include bandwidth, response speed, capture and pull-in range. The LM565 is TTL and DTL compatible. It can produce clean square waves and/or linear triangle waves with in-phase zero crossings.

THE LM567 TONE DECODER

The LM567 is a popular IC that is a special-purpose PLL circuit. Specifically, this device is a tone decoder. It is used in applications in which it is necessary for some circuit to respond to an input signal of a given frequency, but not to signals at other frequencies. The Touch-Tone® system used in telephones is a typical application for tone decoders.

A basic block diagram of the 567's internal circuitry is shown in Fig. 2-5. Notice the similarity to the basic PLL shown earlier. Of course there are some additional stages for the tone decoding function. The pinout diagram for the 8-pin DIP version of the LM567 is shown in Fig. 2-6.

This chip is very sensitive. It can respond to signals with amplitudes as low as a mere 20 mV (0.02 volt) RMS. It can be used to detect signals from 0.01 Hz to 500 kHz (500,000 Hz). The power supply requirements are also flexible. The LM567 can be powered by anything from 4.75 volts to 9 volts. If a 5-volt supply is used, this device will be TTL compatible.

The basic 567 tone decoding circuit is illustrated in Fig. 2-7. The center frequency (frequency to be decoded) is determined by the frequency of the current-controlled oscillator (similar to a VCO), which is set by resistor R1 and capacitor C1. The formula is:

$$F = 1.1/(R1C1)$$

To design a decoding circuit for a specific frequency, simply select a reasonable value for C1, and rearrange the equation to solve for R1:

$$R1 = 1.1/(FC1)$$

For example, let's say we need to detect a 12-kHz (12,000 Hz) signal. If we use a 0.1 μF (0.0000001 farad) capacitor for C1, the resistor should have a value of:

$$R1 = 1.1/(12000 \times 0.0000001$$
$$= 1.1/0.0012$$
$$= 916.6667 \text{ ohms}$$

A 10 kΩ trimpot could be used, and fine tuned for the exact value.

Of course the 567 will also respond to frequencies that are close to the nominal center frequency. Generally, it will respond to anything within 14% of the nominal center frequency. For the 12-kHz decoder we just designed,

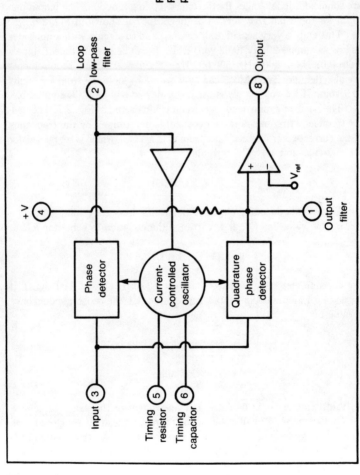

Fig. 2-5. The LM567 is a specialized PLL, designed for tone decoding applications.

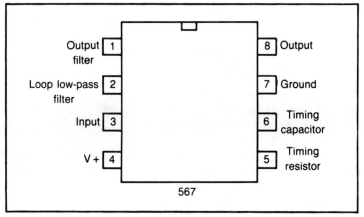

Fig. 2-6. The LM565 is housed in a simple 8-pin DIP housing.

the actual detected frequencies will range from 10.32 kHz to 13.68 kHz.

THE CD4046 DIGITAL PLL

Most PLLs are linear devices, but digital PLLs also exist. The CD4046

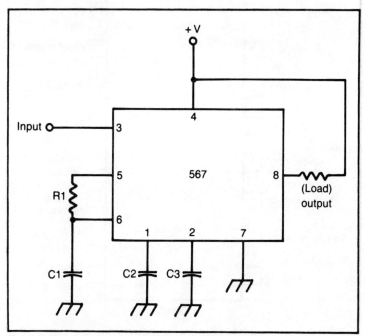

Fig. 2-7. This is the basic LM567 tone decoding circuit.

25

digital PLL is shown in Fig. 2-8. This IC is a CMOS device. Like linear PLLs, the digital PLL is suitable for a great many applications, including:

☐ Frequency modulation and demodulation
☐ Frequency multiplication
☐ Frequency synthesis
☐ FSK demodulation
☐ Tone decoding
☐ Voltage-to-frequency conversion

The internal structure of the CD4046 is illustrated in Fig. 2-9. Notice that this chip contains two phase comparator stages. Phase comparator A is a simple exclusive-OR (X-OR) gate. It provides good noise immunity, but at a price. This comparator tends to lock onto input signals with frequencies close to the harmonics of the VCO frequency. Another disadvantage of this phase comparator is that it will only work if the input signal is a clean square wave with a 50% duty cycle.

Phase comparator B is made up of somewhat more complex circuitry. This comparator is less prone to harmonic lock-up, but its noise immunity

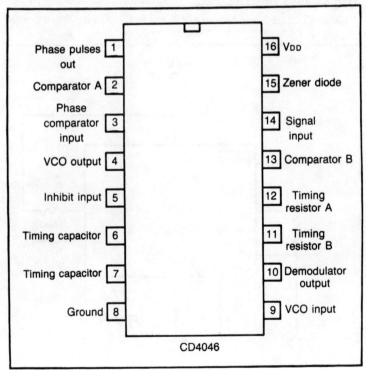

Fig. 2-8. The CD4046 is a digital PLL IC.

26

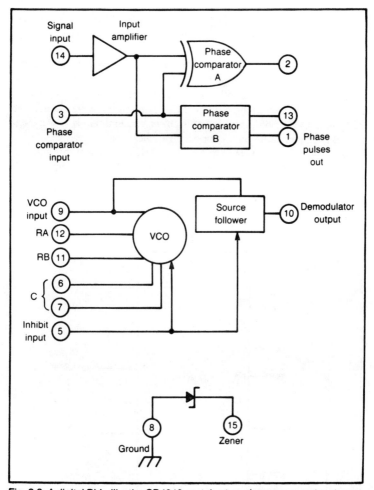

Fig. 2-9. A digital PLL, like the CD4046 contains two phase comparator stages.

isn't as good as for phase comparator A. Comparator B has a wider frequency tracking range than comparator A. Its range is better than 1000:1. In addition, this comparator can function with input pulses with any duty cycle, whereas comparator A is limited to 50% duty-cycle square waves.

Obviously a compromise has to be made, depending on the application. By making both comparators available, the designer is allowed to make the selection himself. If he wants to use phase comparator A, he takes the output off of pin 2. If he wants to use phase comparator B, he takes the output off of pin 13. That's certainly simple enough.

An inhibit input is provided for both the VCO and the source follower. A logic 0 (ground) on this pin enables these stages, while a logic 1 (V_{DD})

turns them off, reducing power consumption. The CD4046 also includes an on-chip 5.2-volt zener diode for voltage regulation applications. Its use is optional.

The power supply for the CD4046 should be between 3 and 18 volts. Current drain is quite low (the exact value will depend on the VCO frequency, and the percentage of time it is enabled via pin 5). Typically, power consumption will be about 1/160th of that required for a standard linear PLL, such as the 565. The CD4046's power consumption is low enough to make it practical in battery-operated applications.

A basic PLL circuit built around the CD4046 is shown in Fig. 2-10. Only a handful of external components are required. Phase comparator A

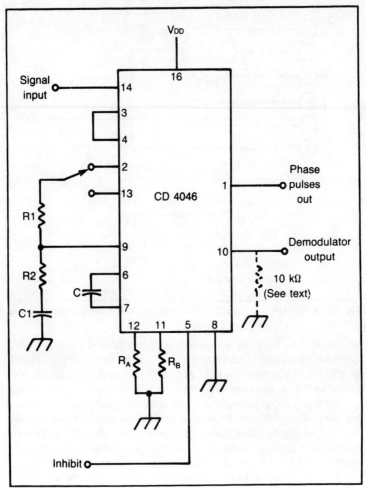

Fig. 2-10. This is a typical PLL circuit using the CD4046.

28

or phase comparator B may be selected with a SPDT switch, as shown here, or one or the other can be permanently connected to R1. Resistors R1 and R2 and capacitor C1 form the loop filter.

The VCO's center frequency is determined by the values of R_a, R_b, and C. In some applications, R_a may be omitted. If only R_b is used, the VCO's center frequency may be varied from 0 Hz (with a 0-volt input) to a maximum value, as defined by this equation:

$$F_{max} = 1/[R_b \times (C + 32 \text{ pF})]$$

This maximum frequency will be generated when the control voltage input is equal to the supply voltage (VDD).

For practical applications, the value of R_b should be between 10 kΩ (10,000 ohms) and 1 megohm (1,000,000 ohms). The capacitor value should be in the 100 pF to 0.01 μF range.

R_a is added to the circuit when a minimum frequency above 0 Hz is required. This component is called an offset resistor, because it offsets the frequency range. The new minimum frequency is defined by this equation:

$$F_{min} = 1/(R_a \times (C + 32 \text{ pF}))$$

The offset maximum frequency is now equal to the sum of the old maximum frequency (as determined in the earlier equation) plus the offset minimum frequency. That is:

$$F_{omax} = F_{max} + F_{min}$$

The acceptable range of values for R_a is the same as for R_b.

If the demodulator output from the source follower (pin 10) is to be used, a load resistor should be connected between this pin and ground. Usually the value of this load resistor will be 10 kΩ (10,000 ohms). If this output is not used in the specific application, pin 10 should be left floating.

PLLS IN TRANSCEIVER APPLICATIONS

Probably the most common class of applications for PLLs in modern electronics is in radio transceivers (transmitters/receivers). They are especially common in CB (Citizens Band) radios, but they are also frequently encountered in FM and television receivers.

PLLs are used for precise tuning. This type of application is called *frequency synthesis*, because a number of channel frequencies can be derived from a single source. For example, a PLL-based CB transceiver can correctly tune in all 40 channels with a single set of crystals. This makes for smaller and less expensive radios, while retaining crystal precision and stability. If a receiver drifts off frequency, it can be an annoying nuisance, but if a transmitter drifts off frequency, it can interfere with other transmissions—which isn't only inconvenient and rude—it's also illegal. Operating an off-frequency transmitter can result in a hefty fine from the FCC.

Figure 2-11 illustrates how a phase-locked loop is used in a frequency-synthesis application. The original reference signal is generated by a crystal oscillator. The use of a crystal here ensures that the reference frequency will be precise and very stable, as it must be for reliable operation.

Next, the reference signal is divided by a specific fixed value (M) in the first counter stage. This gives us a relatively low frequency reference signal (F_r). Typically, this signal will have a frequency of 10 kHz (10,000 Hz).

The 10 kHz reference signal is fed into a PLL (phase detector/filter-amp/VCO). An additional, programmable counter stage is inserted in the feedback loop between the VCO and the phase detector. Because the counter divides the VCO frequency by some value (N), the phase detector will "think" the output frequency is lower than it really is, by a factor of N. As a result, the output will be multiplied by N. In other words, the output frequency will be equal to:

$$F_o = F_r \times N$$

Because the feedback counter stage is programmable, N may take on any of a number of integer values, depending on the information fed to the programming inputs of the counter. This allows the PLL to cover a variety of output values—all using a single crystal. A number of dedicated LSI ICs for frequency synthesis applications have been developed.

The REC86345 Digital Synthesizer IC

Figure 2-12 shows the REC86345 digital synthesizer IC, which is used in a number of commercial CB transceivers. This 18-pin device features a sample-and-hold phase detector which offers very low phase noise output of the VCO. No active filters are required. The *lock detector* output is a very clean digital signal. It is always an unambiguous HIGH or LOW.

Pulldown resistors are included on the chip itself. There is no need to add external pulldown components. The reference oscillator is also included on chip. Only the crystal itself has to be added as an external component. A 10.240 MHz. (10,240,000 Hz) crystal will normally be used. The fixed divider stage can be set up to divide by 1024, or 2048 for a reference frequency (F_r) of either 10 kHz (10,000 Hz) or 5 kHz (5,000 Hz). The selection is made via pin 6. If this pin is fed a logic HIGH signal, the counter will divide the crystal frequency by 2048, giving a reference frequency of 5 kHz. If pin 6 is grounded, or left open (the internal pulldown resistor will make the open pin look to the IC as if it was grounded) the divider will be set up for division by 1024, making the reference frequency equal to 10 kHz.

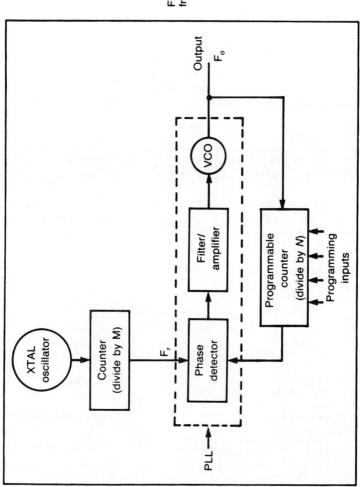

Fig. 2-11. PLLs are often used for frequency synthesis in radio circuits.

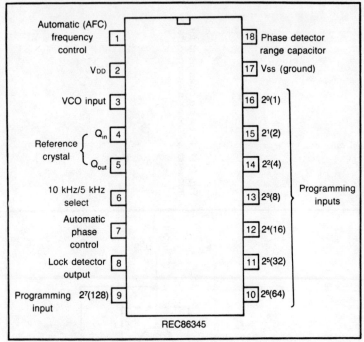

Fig. 2-12. The REC86345 digital synthesizer IC is used in a number of commercial CB transceivers.

Eight binary inputs are used to program the counter/divider, giving 256 possible output frequencies (not all of them will be used in most applications, of course). Table 2-1 shows the division values used for the 40 standard CB channels.

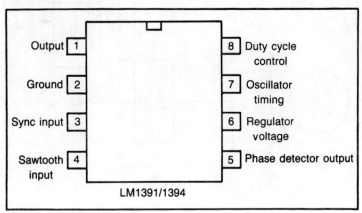

Fig. 2-13. The pinout for the LM1391 and the LM1394 are identical.

Table 2-1. These Are the Division Values
Used for Each of the 40 CB Channels by the REC86345.

Channel	VCO Frequency (MHz)	Divider Value	Pin 9	10	11	12	13	14	15	16
1	37.660	128	1	0	0	0	0	0	0	0
2	37.670	129	1	0	0	0	0	0	0	1
3	37.680	130	1	0	0	0	0	0	1	0
4	37.700	132	1	0	0	0	0	1	0	0
5	37.710	133	1	0	0	0	0	1	0	1
6	37.720	134	1	0	0	0	0	1	1	0
7	37.730	135	1	0	0	0	0	1	1	1
8	37.750	137	1	0	0	0	1	0	0	1
9	37.760	138	1	0	0	0	1	0	1	0
10	37.770	139	1	0	0	0	1	0	1	1
11	37.780	140	1	0	0	0	1	1	0	0
12	37.800	142	1	0	0	0	1	1	1	0
13	37.810	143	1	0	0	0	1	1	1	1
14	37.820	144	1	0	0	1	0	0	0	0
15	37.830	145	1	0	0	1	0	0	0	1
16	37.850	147	1	0	0	1	0	0	1	1
17	37.860	148	1	0	0	1	0	1	0	0
18	37.870	149	1	0	0	1	0	1	0	1
19	37.880	150	1	0	0	1	0	1	1	0
20	37.900	152	1	0	0	1	1	0	0	0
21	37.910	153	1	0	0	1	1	0	0	1
22	37.920	154	1	0	0	1	1	0	1	0
23	37.950	157	1	0	0	1	1	1	0	1
24	37.930	155	1	0	0	1	1	0	1	1
25	37.940	156	1	0	0	1	1	1	0	0
26	37.960	158	1	0	0	1	1	1	1	0
27	37.970	159	1	0	0	1	1	1	1	1
28	37.980	160	1	0	1	0	0	0	0	0
29	37.990	161	1	0	1	0	0	0	0	1
30	38.000	162	1	0	1	0	0	0	1	0
31	38.010	163	1	0	1	0	0	0	1	1
32	38.020	164	1	0	1	0	0	1	0	0
33	38.030	165	1	0	1	0	0	1	0	1
34	38.040	166	1	0	1	0	0	1	1	0
35	38.060	167	1	0	1	0	0	1	1	1
36	38.050	168	1	0	1	0	1	0	0	0
37	38.070	169	1	0	1	0	1	0	0	1
38	38.080	170	1	0	1	0	1	0	1	0
39	38.090	171	1	0	1	0	1	0	1	1
40	38.100	172	1	0	1	0	1	1	0	0

The REC86345 consumes very little power. The power dissipation is typically less than 10 mW (0.01 watt). Pin 1 is an *automatic frequency control* (AFC) output that is used to bring the VCO frequency with the lock range of the phase detector. This is a three-state output. It is open when the circuit is properly phase locked. If the VCO frequency is too low, positive-going pulses will appear at pin 1. Negative going pulses at this pin indicate that the VCO frequency is too high.

A true digital indicator of whether the circuit is phase locked or not is offered at pin 8. If the VCO frequency is off (either too low or too high) pin 8 will go logic LOW. If the circuit is properly phase locked, a logic HIGH signal will appear at pin 8. The lock detector output will go LOW whenever an error condition exists for more than 0.5 ms (0.0005 second). This signal is often used to inhibit transmitter operation if there is any problem in locking-on the desired frequency.

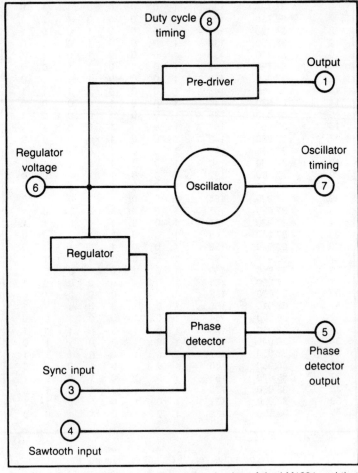

Fig. 2-14. The only difference between the circuitry of the LM1391 and that of the LM1394 is the polarity of the phase detector.

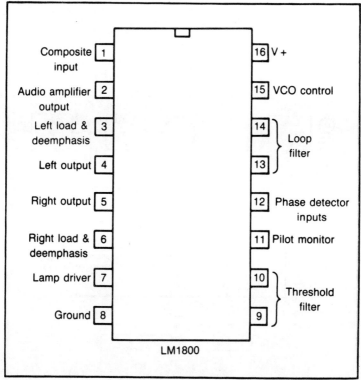

Fig. 2-15. The LM1800 is a PLL IC designed for demodulating stereo FM broadcasts.

THE LM1391 AND LM1394 PLL BLOCKS

Phase-locked loops are also sometimes found in the horizontal section of television receivers. Two chips that were designed with this application in mind are the LM1391 and 1394, shown in Fig. 2-13. The pinout is the same for both of these ICs. The basic internal structure is illustrated in Fig. 2-14. The only difference between these two units is the polarity of the phase detector. While these devices were designed for the horizontal circuits in a TV set, they can be used in other relatively low frequency signal processing applications.

THE LM1800 PLL FM STEREO DEMODULATOR IC

Another frequent application for phase-locked loops is for demodulation of stereo FM broadcasts. The PLL is used to regenerate the 38 kHz subcarrier. A special-purpose IC for this application is the LM1800, which is shown in Fig. 2-15. The internal circuitry is illustrated in Fig. 2-16. This chip produces high fidelity sound at low cost, making it suitable for use

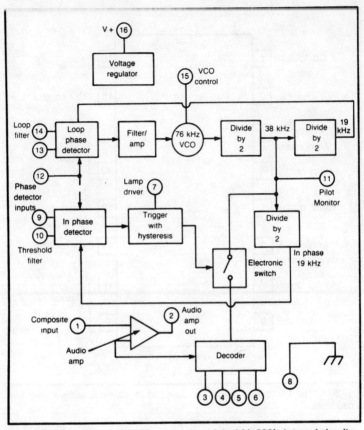

Fig. 2-16. This is a functional block diagram of the LM1800's internal circuitry.

in inexpensive stereo receivers. Unlike most older radio circuits, no tuning coils are needed. All tuning can be accomplished with a single potentiometer. The LM1800 can also automatically switch between monaural and stereophonic operation.

Chapter 3

Timers

MANY ELECTRONICS APPLICATIONS REQUIRE SOME KIND OF TIMING signal to keep various stages in synchronization, or to trigger events periodically, or at a specific time. Not surprisingly, a circuit that handles such timing functions is called a timer. Timers are among the most popular types of ICs because they are so easy to work with and have so many applications.

There are two basic types of timer circuits. Both are *multivibrator* (square wave) *circuits*. The *monostable multivibrator* has one stable state (HIGH or LOW) that it will normally remain in until a triggering signal is received. Upon being triggered, the output switches to the opposite (unstable) state for a fixed period of time (regardless of the length of the trigger pulse). At the end of the timing cycle, the output reverts to its normal, stable state. The action of a monostable multivibrator is illustrated in Fig. 3-1. This type of circuit is sometimes called a *one-shot*.

The other common type of timer circuit is the *astable multivibrator*. This circuit has no stable states. Its output continuously switches back and forth between states (HIGH and LOW) at a regular rate, as shown in Fig. 3-2. It is nothing more nor less than a square-wave generator. When used for timing purposes, this type of circuit is often referred to as a *clock*.

Bistable multivibrator circuits (with two stable output states) also exist, but they are not used for timing purposes, so we will not consider them here.

THE 555 TIMER

One of the most popular integrated circuits of all time is the 555 timer. This readily available and inexpensive chip can be used in countless tim-

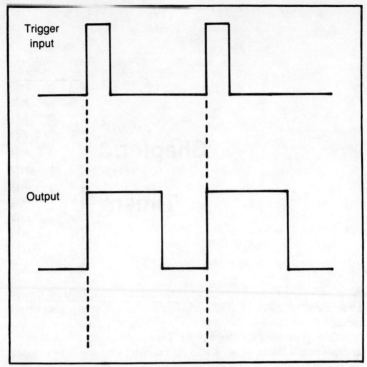

Fig. 3-1. A monostable multivibrator has a single stable state, which it reverts to after its time period.

ing applications. It offers high reliability and reasonable precision at low-cost. Designs involving the 555 are usually fairly simple and straightforward.

The 555 is normally supplied in an 8-pin DIP case, as shown in Fig. 3-3. The basic internal structure of this handy device is illustrated in Fig. 3-4. The 555 timer may be used in either linear or digital circuits. It is a linear device itself, based on bipolar transistors. It can be operated from a wide range of supply voltages. Anything from +4.5 volts to +18 volts will do. In most applications, +9 volts to +12 volts will give the best performance. If the 555 is to be used with digital ICs, it should use the same

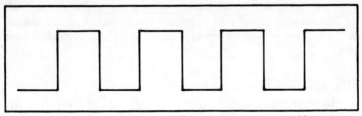

Fig. 3-2. An astable multivibrator oscillates between two unstable states.

38

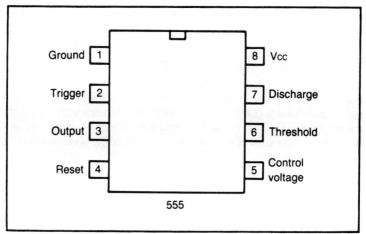

Fig. 3-3. The 555 timer IC is one of the most popular ICs around.

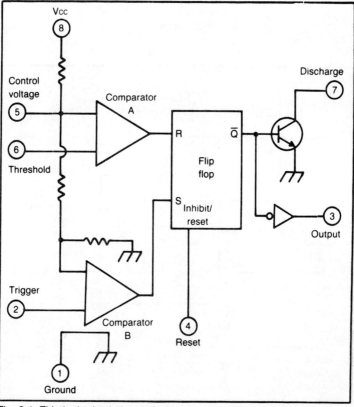

Fig. 3-4. This is the basic internal structure of the 555 timer IC.

39

supply voltage — for example, +5 volts for TTL devices. This versatile IC can be used for either monostable or astable applications.

The 555 Monostable Multivibrator

The basic 555 monostable circuit is illustrated in Fig. 3-5. Notice how few external components are required. The normal stable state for the output of this circuit is LOW (near ground potential). When a negative-going trigger pulse is received, the output snaps high for a fixed time period determined by the values of resistor R1 and capacitor C1, according to this simple formula:

$$T = 1.1R1C1$$

For example, if R1 has a value of 39 kΩ (39,000 ohms) and C1 has a value of 0.022 μF (0.000000022 farad), every time the circuit is triggered, it will put out a HIGH pulse that lasts:

$$T = 1.1 \times 39000 \times 0.000000022$$
$$= 0.0009438 \text{ second}$$
$$= 0.001 \text{ second}$$

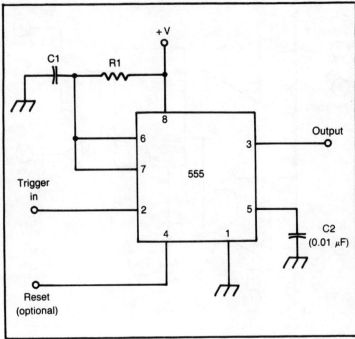

Fig. 3-5. The 555 timer IC can be used in a monostable multivibrator circuit.

40

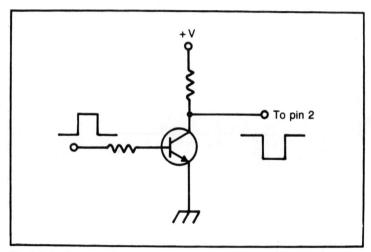

Fig. 3-6. In some applications it may be necessary to invert the trigger signal before applying it to a 555 timer.

The 555 responds to a negative-going trigger pulse. In some applications we may have a positive-going pulse that we want to use to trigger a timer. Figure 3-6 shows a simple trigger inverter circuit that can be used with the 555.

The 555 Astable Multivibrator

The basic 555 astable multivibrator circuit is shown in Fig. 3-7. Notice how similar it is to the monostable multivibrator circuit presented earlier. There are two times to calculate in this circuit. The time the output is in the HIGH state, and the time it is in the LOW state. The complete cycle time is the sum of the HIGH and LOW times. The HIGH time is defined by C1 and both resistors, according to this formula:

$$Th = 0.693 \ C1(R1 + R2)$$

For example, if we assume the following component values:

$$R1 = 22\Omega \ (22,000 \text{ ohms})$$
$$R2 = 47\Omega \ (47,000 \text{ ohms})$$
$$C1 = 0.33\mu F \ (0.00000033 \text{ farad})$$

the HIGH time would be:

$$Th = 0.693 \times 0.00000033 \times (22000 + 47000)$$
$$= 0.0000022869 \times 69000$$

41

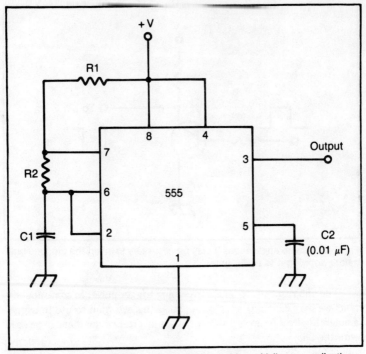

Fig. 3-7. The 555 timer IC can also be used in astable multivibrator applications.

$$= 0.01577961 \text{ second}$$
$$= 0.016 \text{ second}$$

For the LOW time, however, resistor R1 is ignored, so the formula for determining this part of the cycle is simply:

$$T1 = 0.693 \; C1R2$$

For our example, this works out to:

$$T1 = 0.693 \times 0.00000033 \times 47000$$
$$= 0.01074843 \text{ second}$$
$$= 0.011 \text{ second}$$

The total cycle time is simply the sum of the HIGH time, and the LOW time, or in our example:

$$T = Th + Tl$$
$$= 0.016 + 0.011$$
$$= 0.027 \text{ second}$$

It is usually more practical to talk about a waveform in terms of frequency instead of cycle time. The frequency is simply the reciprocal of the cycle time. That is:

$$F = 1/T$$
$$= 1/(Th + Tl)$$
$$= 1/[(0.693 \text{ C}1 \text{ (R1 + R2)}] + (0.693\text{C1R2})$$
$$= 1/[0.693\text{C}1 \text{ (R1 + 2R2)}]$$

Using the component values from our example, we get an output signal frequency of:

$$F = 1/(0693 \times 0.00000033 \times [22000 + (2 \times 47000]$$
$$= 1/(0.0000022869 \times 116000)$$
$$= 1/0.02652804$$
$$= 38 \text{ Hz}.$$

The 555 Dual Timer

The 555 timer is such a handy and versatile device that in many cir-

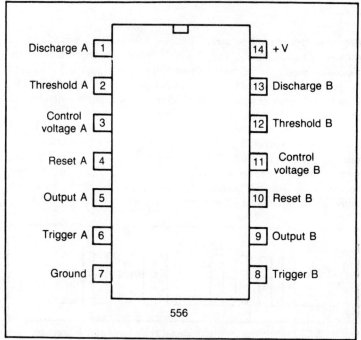

Fig. 3-8. The 556 dual timer is the equivalent of two 555s on a single chip.

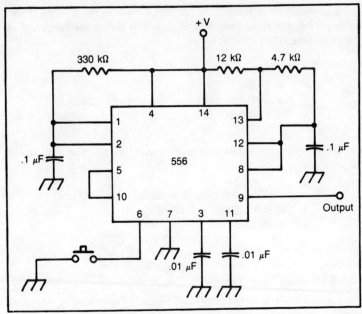

Fig. 3-9. This tone-burst generator circuit is a typical application for the 556 dual timer.

cuits more than one will be used. When this is the case, it is generally more convenient to use the 556 dual timer IC. This device contains two complete and independent 555 type timers in a single 14-pin package. The pinout diagram for the 556 dual timer IC is shown in Fig. 3-8. Half a 556 can be used in any application for a 555. Of course, you need to watch out for the differences in the pin numbers.

Figure 3-9 shows a typical application for the 556 dual timer. This is a tone-burst generator. The output will be regular bursts of a tone, separated by periods of silence. The waveform is illustrated in Fig. 3-10. The same circuit using two separate 555's is shown in Fig. 3-11. Timer B generates the tone. Timer A generates a slower rectangle wave, which turns timer B's output on and off at regular intervals.

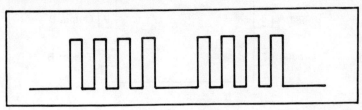

Fig. 3-10. The tone-burst generator circuit shown in Fig. 3-9 produces this type of signal.

44

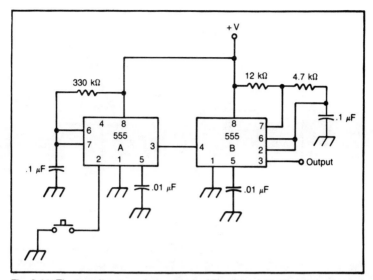

Fig. 3-11. The tone-burst generator circuit of Fig. 3-9 could also be constructed with two separate 555s.

Another frequent application for a 556 dual timer (or a pair of 555 timers) is to increase the time period of a monostable multivibrator. A single 555 timer can generate output pulses ranging from a fraction of a second to several minutes. Ten or fifteen minutes is about the maximum unless a special low-leakage capacitor is used for C1. Such capacitors are hard to find and very expensive. It is often more practical to achieve longer time periods by cascading two 555 monostable stages, as illustrated in Fig. 3-12. The same circuit using a single 556 instead of a pair of 555's is shown in Fig. 3-13.

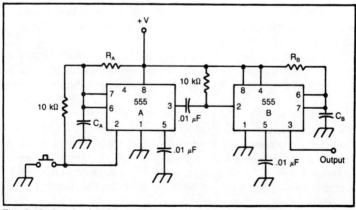

Fig. 3-12. A pair of 555s could be cascaded to create a longer time period.

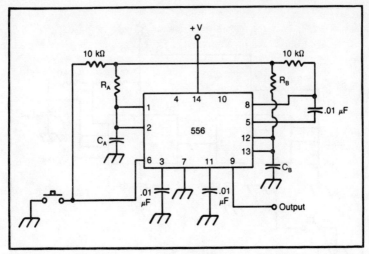

Fig. 3-13. The circuit shown in Fig. 3-12 can be simplified by using a single 556 dual timer IC.

The 558 Quad Timer

A quad timer IC is also available. This is the 558. It contains four 555 type timer stages, but it is not suitable for all applications, because not all functions are brought out to the IC pins as with the 555 and 556. A pinout diagram for this device is shown in Fig. 3-14.

WIDE RANGE PRECISION MONOSTABLE TIMERS

As useful and versatile as the 555 timer IC is, it is unquestionably less than perfect. The time periods cannot be determined with complete precision. Moreover, the maximum possible time period is severely limited. In the vast majority of timing applications, a 555 will probably do just fine, and there's no reason not to use it. But in some applications something better is required.

Several improved timer ICs have been developed to allow for greater precision, and longer time periods. Two devices in this category are the 322 and the 3905. Both of these ICs are wide-range precision monostable timers.

In the real world compromises are inevitable. To achieve the improvements in the 322 and 3905, something had to be given up. These devices can be used in monostable applications only. They cannot be used for astable operation. This is less of a shortcoming than it might seem. The special advantages of the 322 and 3905 (greater precision and longer time periods) generally are only needed in monostable circuits. They are rarely relevant for astable multivibrators circuits.

There are other advantages to the 322 and 3905 too. They feature greater immunity to noise from the power supply than the 555. These devices can be operated from a wide range of supply voltages (from +4.5 volts to +40 volts). An on-chip voltage regulator is used as a reference for the timing comparators. This gives greater accuracy in determining the time period, because the supply voltage has virtually no effect, even if it changes during the timing period.

The output stage of the 322 and 3905 can produce higher voltages than that of the 555. The output stage in these improved units is also more generally flexible and versatile. It can be wired in either common-collector or common-emitter configurations.

Another versatile feature of these ICs is that the stable output state may be externally selected with the "Logic" pin. If this pin is made HIGH, the output will normally be LOW (HIGH during the timing period). Of course, the reverse holds true when the Logic input is held LOW. An additional advantage is that the timing period formula is as simple as it can possibly be:

$$T = RC$$

No constant is required, as with the 555.

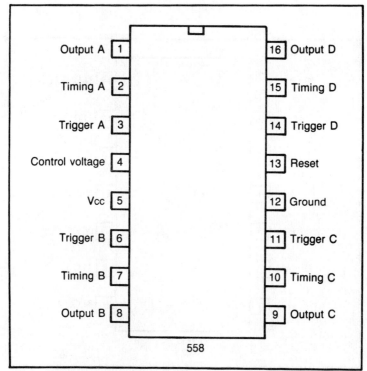

Fig. 3-14. The 558 is a quad 555 timer IC.

These timers can be set up for timing periods of up to several hours. The only limitation is the value of the timing components. The pinout diagram for the 322 is shown in Fig. 3-15, and the 3905 is shown in Fig. 3-16. A basic block diagram of both devices is shown in Fig. 3-17. These two devices are very similar, and will be used in essentially the same circuits. For convenience, we will just show the pin numbers for the 322.

A basic monostable multivibrator circuit built around the 322 timer is illustrated in Fig. 3-18. The output is in the common-collector mode in this circuit. A similar circuit with a common-emitter output is shown in Fig. 3-19.

PROGRAMMABLE TIMERS

Even greater versatility can be achieved with a programmable timer. A typical device of this type is the XR2240. As the block diagram of Fig. 3-20 shows, this IC contains an 8-bit binary counter along with the actual timer circuitry. The pinout diagram for this device is given in Fig. 3-21. With this device 255 different time periods can be achieved with a single resistor/capacitor combination. They are all integer multiples of the basic time period, which is equal to:

$$T = RC$$

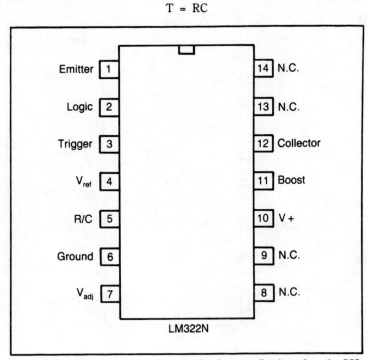

Fig. 3-15. The 322 is used for more precise timer applications than the 555.

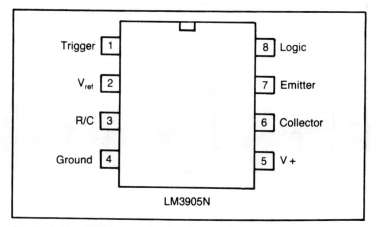

Fig. 3-16. The 3905 is very closely related to the 322.

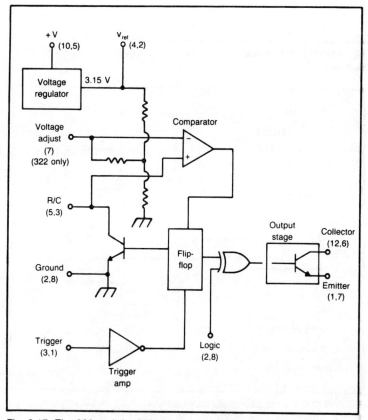

Fig. 3-17. The 322 and the 3905 have very similar internal circuitry.

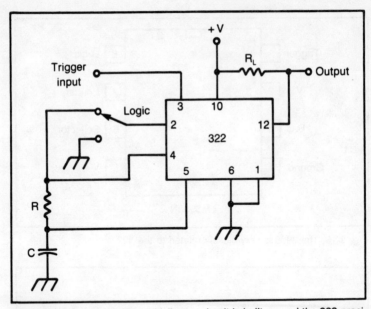

Fig. 3-18. This monostable multivibrator circuit is built around the 322 precision timer IC.

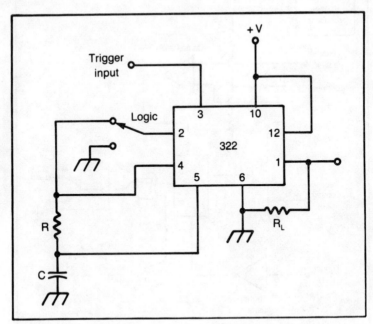

Fig. 3-19. This is a variation on the circuit shown in Fig. 3-18, but with a common-emitter output.

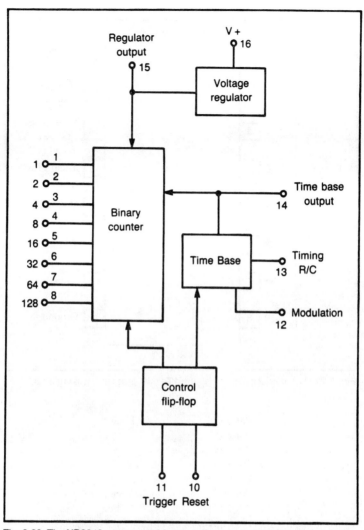

Fig. 3-20. The XR2240 programmable timer IC contains an 8-bit binary counter, along with the basic timer circuitry.

For example, if R = 1 MΩ (1,000,000 ohms) and C = 0.1 μF (0.0000001 farad), the basic time period would be:

$$T = 1000000 \times 0.0000001$$
$$= 0.1 \text{ second}$$

At each of the counter output pins we would find the following time periods:

51

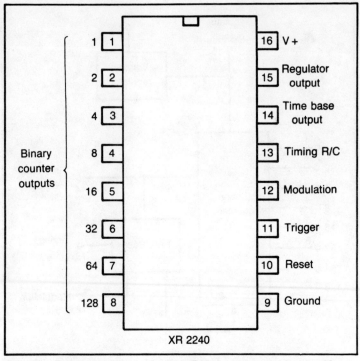

Fig. 3-21. The pinout diagram for the XR2240 programmable timer IC.

Pin 1 (1)	0.1 Sec
Pin 2 (2)	0.2 Sec
Pin 3 (4)	0.4 Sec
Pin 4 (8)	0.8 Sec
Pin 5 (16)	1.6 Sec
Pin 6 (32)	3.2 Sec
Pin 7 (64)	6.4 Sec
Pin 8 (128)	12.8 Sec

These outputs can also be combined for intermediate values. For instance, if we combine the outputs from pins 2, 5, and 7, we will get a time period of:

$$T_x = 2T + 5T + 7T$$
$$= 0.2 + 1.6 + 6.4$$
$$= 8.2 \text{ seconds}$$

The counter outputs can be used to create some very long periods. For example, let's say $R = 1 \text{ M}\Omega$ (1,000,000 ohms), and $C = 500 \mu F$ (0.0005 farad). These are about the maximum values we can use with reasonable confidence of fair stability.

With these component values, the basic time period works out to:

$$T = 1000000 \times 0.0005$$
$$= 5000 \text{ seconds}$$
$$= 83 \text{ minutes, } 20 \text{ seconds}$$
$$= 1 \text{ hour, } 23 \text{ minutes, } 20 \text{ seconds}$$

For convenience in our discussion, we will simply round this off to 1.5 hours. This is about the maximum time period we can hope for from a 322 or 3905.

Using the binary counter of the XR2240 we can get time periods that are much longer:

Pin 1 (1)	1.5 Hrs
Pin 2 (2)	3 Hrs
Pin 3 (4)	6 Hrs
Pin 4 (8)	12 Hrs
Pin 5 (16)	24 Hrs
Pin 6 (32)	48 Hrs
Pin 7 (64)	96 Hrs
Pin 8 (128)	192 Hrs

By combining all eight of the binary counter outputs (for a total multiplier of 255), the maximum time period becomes:

$$T_{max} = T \times 255$$
$$= 1.5 \times 255$$
$$= 382.5 \text{ Hours}$$
$$= 15 \text{ Days, } 22.5 \text{ Hours}$$

Now, how often are you likely to need a time period longer than that?

The XR2240 Monostable Circuit

The basic XR2240 monostable circuit is illustrated in Fig. 3-22. The output will have a time period of:

$$T = nRC$$

where n is the programmed value (from 1 to 255) set up by pins 1 through 8, R is the value of the timing resistor (1 kΩ to 10 MΩ), and C is the value of the timing capacitor (0.01 μF to 1000 μ). These limits mean the time period can be as short as:

$$N = 1$$

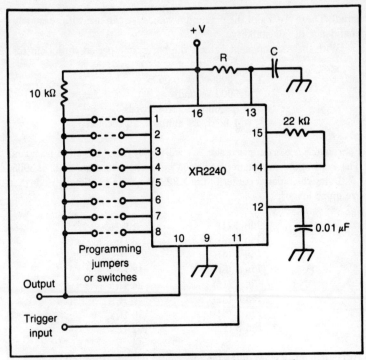

Fig. 3-22. The basic XR2240 monostable multivibrator circuit.

$$R = 1 \text{ k}\Omega = 1000 \text{ ohms}$$
$$C = 0.01 \text{ }\mu F = 0.00000001 \text{ farad}$$
$$T = 1 \times 1000 \times 0.00000001$$
$$= 0.00001 \text{ second}$$

or as long as:

$$N = 255$$
$$R = 10 \text{ M}\Omega = 10,000,000 \text{ ohms}$$
$$C = 1000 \text{ }\mu F = 0.001 \text{ farad}$$
$$T = 255 \times 10000000 \times 0.001$$
$$= 2,550,000 \text{ seconds}$$
$$= 42,500 \text{ minutes}$$
$$= 708 \text{ hours, 20 minutes}$$
$$= 29 \text{ days, 12 hours, 20 minutes}$$

Quite an impressive range! A positive going pulse is used to trigger the XR2240.

54

This IC can be powered with a fairly wide range of supply voltages. The circuit will operate correctly if powered by anything from +5 volts to +15 volts. If the supply voltage is greater than, or equal to +7 volts, and the timing capacitor is less than or equal to 0.1 μF, a shunt capacitor should be connected between the time base output (pin 14) and ground. This shunt capacitor should have a value between 250 and 300 pF.

The XR2240 Astable Circuit

The XR2240 astable multivibrator is quite similar to the monostable multivibrator circuit we just discussed. The astable version is shown in Fig. 3-23. The primary difference between this astable circuit and the monostable version is the reset input (pin 10) is not connected to the output. The circuit may be reset with an external signal, if desired, otherwise, this pin is simply ignored. The output frequency can be determined with this formula:

$$F = 1/(2nRC)$$

where n is the program value (1 to 255) set up by pins 1 through 8, R is

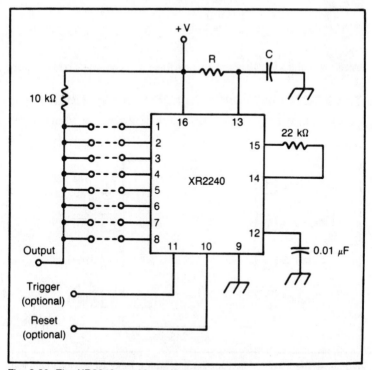

Fig. 3-23. The XR2240 astable multivibrator circuit.

the timing resistance (1 kΩ to 10 MΩ), and C is the timing capacitance (0.01 µF to 1000 µF).

The minimum frequency is achieved when all three values are at their maximums:

$$N = 255$$
$$R = 10 \text{ M}\Omega = 10,000,000 \text{ ohms}$$
$$C = 1000 \text{ }\mu\text{F} = 0.001 \text{ farad}$$
$$F = 1/(2 \times 255 \times 10000000 \times 0.001)$$
$$= 1/5100000$$
$$= 0.0000002 \text{ Hz}.$$

The maximum frequency can be obtained by reducing all three variables to their minimum values:

$$N = 1$$
$$R = 1 \text{ k}\Omega = 1000 \text{ ohms}$$
$$C = 0.01 \text{ }\mu\text{F} = 0.00000001 \text{ farad}$$
$$F = 1/(2 \times 1 \times 1000 \times 0.00000001)$$
$$= 1/0.00002$$
$$= 50,000 \text{ Hz}$$
$$= 50 \text{ kHz}.$$

Once again, the XR2240 programmable timer gives us a remarkable range.

THE XR-2243 MICROPOWER LONG RANGE TIMER IC

Currently the 555 is the unquestioned king of timer ICs when it comes

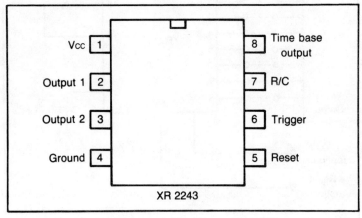

Fig. 3-24. The XR2243 Micropower Long Range Timer IC is used in applications which require very long time periods.

to popularity. The 555's reign might not last much longer, however. Many new timer ICs that are as easy to work with (if not even easier) than the 555 are appearing on the market. These new devices offer several significant advantages over the older 555. On top of everything else, the cost of many such chips are competitive with the 555. All in all, it's looking like there is less and less reason for sticking with the old, reliable 555.

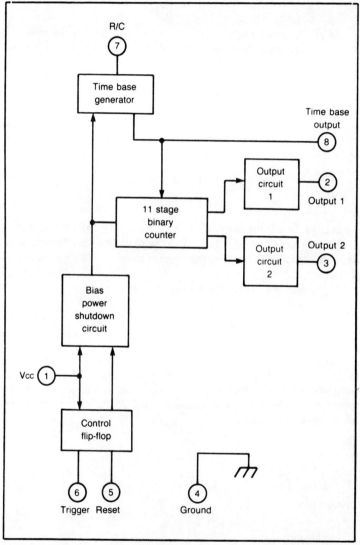

Fig. 3-25. This is the basic internal circuitry of the XR2243 Micropower Long Range Timer IC.

One such device is the XR-2243 Micropower Long Range Timer IC, which is shown in Fig. 3-24. The basic internal circuitry of this chip is shown in Fig. 3-25. This monolithic timer can generate time periods ranging from microseconds to several days. The long range time periods are the result of an internal 11-stage binary counter. At pin 8 the basic time period is available. The formula for the time period is simply:

$$T = RC$$

At pin 3 a much longer output pulse appears. This is the result of passing the basic time-base signal through the internal counter. The time period at this output is equal to:

$$T1 = 1024 \ RC$$

In addition, two or more XR-2243's may be cascaded to create even longer time periods. With two XR-2243's in series, the time period can be as long as:

$$T12 = 1024 \times 1024 \ RC$$
$$= 1048576 \ RC$$

A third XR-2243 would give a time period equal to 1024 cubed, multiplied by the basic time-base value (RC). By cascading XR-2243 stages, a designer can easily generate time periods of days, months, or even years, if necessary.

A square wave signal with a frequency of:

$$F = 1/2 \ RC$$

is available at pin 2.

Besides the long time periods available, this chip has another important advantage — it doesn't consume much power. In normal operation it consumes less than 1 mA (0.001 amp). On standby, the current drain is less than 100 μA (0.0001). This makes the XR-2243 a perfect choice for battery-powered equipment.

```
         ┌──────U──────┐
       ╓─┤ 1    S    24 ├─╖
       ╓─┤ 2    P    23 ├─╖
       ╓─┤ 3    E    22 ├─╖
       ╓─┤ 4    C    21 ├─╖
       ╓─┤ 5    I    20 ├─╖
       ╓─┤ 6    A    19 ├─╖
       ╓─┤ 7    L    18 ├─╖
       ╓─┤ 8    P    17 ├─╖
       ╓─┤ 9    U    16 ├─╖
       ╓─┤10    R    15 ├─╖
       ╓─┤11    P    14 ├─╖
       ╓─┤12    O    13 ├─╖
              S
              E
```

Chapter 4

LED Drivers

M OST ELECTRONIC CIRCUITS WON'T DO ANYONE MUCH GOOD AT ALL if there isn't some way for it to indicate what's going on within the circuitry. Some kind of output display is needed. One of the most commonly used display devices in modern electronics is the LED *(light-emitting diode)*. It is small, convenient, and inexpensive. Moreover, it can be used to make an effective and eye-catching display.

An LED is simply a special diode that glows when it is forward biased. Usually they are red, but other colors, such as green and yellow, are also available. Infrared LEDs are also manufactured for special purposes.

LEDs may be used individually, or in grouped packages. Bargraph LEDs have several independent LEDs with a common element (either the anodes or the cathodes) internally tied together. A bargraph LED array is shown in Fig. 4-1.

Another common packaging of multiple LEDs is the seven-segment display. Seven dash-like LEDs are arranged in a figure-8 pattern, with a common element (common-anode or common-cathode) as illustrated in Fig. 4-2. Any of the 10 basic digits (0 - 9) and many letters can be displayed by activating the proper combination of LED segments. Many seven-segment displays also include an eighth segment to represent a decimal point.

THE THREE-STATE LED

A three-state LED can glow with any of three different colors, depending on how the voltage is applied. This device is not really an integrated circuit, but it is closely enough related to warrant a mention here. It is not a complex device at all. A three-state LED is nothing more than a pair of

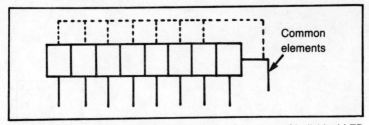

Fig. 4-1. Bargraph LED displays are made up of an array of individual LED elements.

back-to-back LEDs in a single housing, as illustrated in Fig. 4-3.

When a dc voltage is applied to a three-state LED, one of the internal diodes will be forward biased, and the other will be reverse biased. That is, one will be lit, and the other will be dark. If the dc polarity is reversed, the states of the LEDs will also be reversed. The internal diodes glow in different colors. Usually one is red, and the other is green.

If an ac signal is applied to a three-state LED, the two internal LEDs will alternately blink on and off. If the ac frequency is above about 10 Hz, the human eye will be unable to distinquish the individual flashes. Instead, both internal LEDs will appear to be continuously lit. Their different colors will blend together to produce a third color—typically yellow. Thus, the three-state LED gives a clear, unambiguous indication of three different electrical conditions:

□ Positive dc   Red
□ Negative dc   Green
□ Ac   Yellow

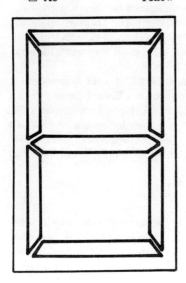

Fig. 4-2. Seven LED segments arranged in a figure-8 pattern can be selectively lit to display any digit, and some letters.

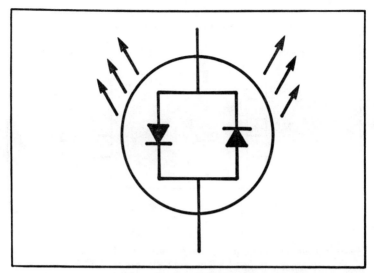

Fig. 4-3. A three-state LED is made up of two back-to-back LEDs in a single housing.

Clearly, the three-state LED can be an immensely useful indicator device.

THE LM3909 LED FLASHER/OSCILLATOR

The LM3909 is an external low-frequency oscillator IC, which is designed to flash an LED on and off. The pinout diagram for this simple 8-pin chip is shown in Fig. 4-4. Only two external components are required—the LED itself, and a capacitor which determines the flash rate. The circuit can be operated with a simple 1.5-volt battery. The supply voltage can range from 1.15 volts to 6.0 volts. The low current drain of this circuit (typically under 0.5 mA (0.0005 amp) also makes it very suitable for battery-powered circuits.

The LM3909 outputs pulses of about 2 volts to the LED, even if the supply voltage is only 1.5 volts. The oscillator is inherently self-starting. No external signal pulse is required. The basic LED flasher circuit built around the LM3909 is illustrated in Fig. 4-5.

BARGRAPH DRIVER ICs

A number of ICs designed for creating LED bargraphs have been put on the market over the years. A *bargraph* is simply a series of LEDs, usually arranged in a straight line, as shown in Fig. 4-1. Other configurations (such as arcs) might also be used in some special applications. The driving circuitry is set up so that each LED is lit when the input voltage exceeds a

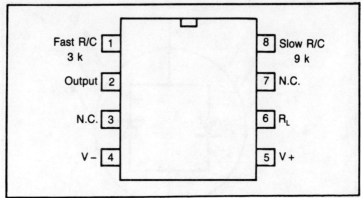

Fig. 4-4. The LM3909 is a low-frequency oscillator designed for flashing an LED on and off.

specific level. Therefore, the bargraph indicates the approximate magnitude of the input voltage by how many LEDs are lit. All LEDs with turn-on points below the input voltage will be on. All LEDs with turn-on points above the input voltage will be off.

A bargraph driving circuit is made up of a voltage-divider network, and a series of comparators, as illustrated in Fig. 4-6. Usually some kind of drive circuitry to supply the necessary current to the LEDs is also included. The display may respond either linearly or logarithmically, depending on the scaling used in the reference voltage network (voltage divider).

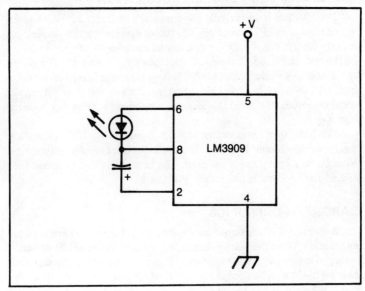

Fig. 4-5. This simple LED flasher circuit is built around the LM3909.

Individual LEDs may be used to make up the bargraph, but a dedicated display unit containing multiple LEDs in a single package usually looks a little better, and is more compact. In addition, the internal LEDs in a multiple display unit are matched for color and brightness.

Several bargraph display units are available. For example, Hewlett-Packard manufactures the HDSP-4820, HDSP-4830, and HDSP-4840. Each of these units contains a row of ten matched LEDs. The HDSP-4820 is standard red, the HDSP-4830 is high-efficiency red, and the HDSP-4840 is high-efficiency yellow. Other bargraph display units are also made. Some have fewer LEDs, and others have more.

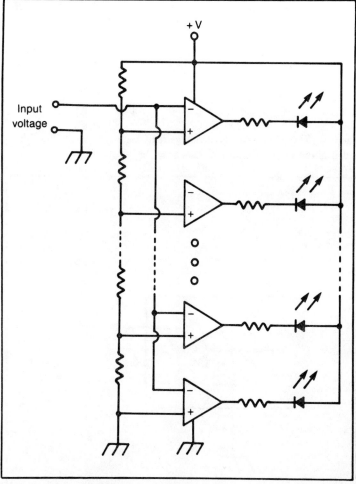

Fig. 4-6. A simple bargraph driver circuit is made up of a voltage divider network and a series of comparators.

The TL490C And TL491C Bargraph Driver ICs

A pair of typical bargraph driver ICs are the TL490C and the TL491C. These two devices are essentially identical, except for the configuration of the outputs. The TL490C has open collector outputs, as shown in Fig. 4-7. These outputs can sink up to 40 mA (0.04 amp) at 32-volts maximum. The TL491C, on the other hand, features open emitter outputs, as illustrated in Fig. 4-8. These outputs can source up to 25 mA (0.025 amp) at 55 volts maximum.

The TL490C and the TL491C are the same except for the output configuration. We will just discuss the TL490C. Just remember that everything we say here applies to the TL491C too.

The TL490C is a ten-step analog level (voltage) detector. It will light a ten-element row of LEDs. A THRESHOLD input is provided for adjustable sensitivity. The increments between adjacent LEDs is adjustable from 50 mV (0.05 volt) to 200 mV (0.2 volt) per LED. The input voltage required to light the first (lowest level) LED can be calculated via the following formula:

$$V_{in} = 0.84/\{1 + [2240(R + 700)]/(700R)\}$$

where V_{in} is the input threshold voltage required to turn on the first LED, and R is the value of a resistor connected between pin 6 and ground.

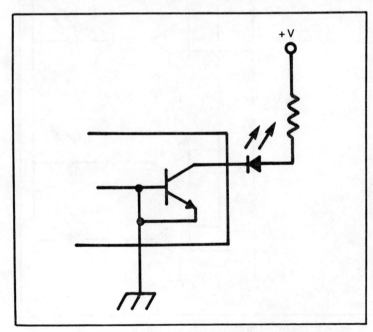

Fig. 4-7. The TL490C bargraph driver IC features open-collector outputs.

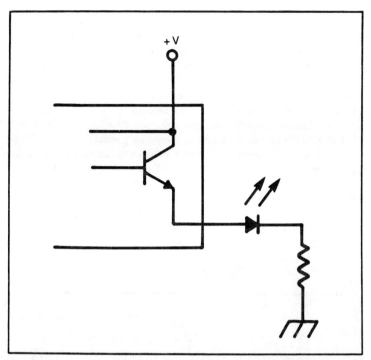

Fig. 4-8. The TL491C is similar to the TL490C, but it has open-emitter outputs.

Alternatively, the value of R can be determined experimentally. Use a 1 kΩ (1000 ohms) trimmer potentiometer for R. Feed in the voltage you want to turn on the first LED. Then simply adjust the trimpot until the LED just lights up.

A basic bargraph circuit using the TL490C is shown in Fig. 4-9. Most of the work is done within the IC itself. Only R, the LEDs and their appropriate current-limiting resistors are required as external components.

The LM3914 Dot/Bar Display Driver IC

One of the most versatile bargraph display driver ICs around is the LM3914. Like the TL490C and TL491C, it contains a voltage divider and ten comparators. It can drive a ten-LED bargraph display.

One of the several advantages that the LM3914 offers over most other bargraph display drivers is that a single resistor is used for current limiting for all of the LEDs. An individual resistor is not needed for each individual LED in the display. This cuts down circuit cost somewhat, and allows for a more compact circuit.

A typical bargraph circuit built around the LM3914 is shown in Fig. 4-10. One of the unique features of this IC is the mode control. If pin 9

65

is connected to the supply voltage (+ Vcc—pin 3) the circuit will function as a standard bargraph display. All LEDs below the input value will be lit.

Alternatively, pin 9 can be connected to pin 11, as illustrated in Fig. 4-11. Now the circuit will function as a moving dot display. Only a single LED indicating the current input voltage will be lit. All other LEDS (both lower and higher) will be dark. In some applications this will give a clearer indication. A moving dot display can also be achieved by leaving pin 9 floating (not connected to anything). Moving dot display drivers are usually not at all easy to design, so this is a particularly handy feature of the LM3914.

A moving dot display also offers a number of additional fascinating applications. For example, it could be used as the heart of a solid-state oscilloscope with a LED screen. Other applications can arise from thinking of the circuit as a one-of-ten output selector. The outputs do not necessarily have to be LEDs. The circuit could drive low power relays, drive transistors, optoisolators, or SCRs. One of ten circuits can be activated, depending on the present value of the input signal.

The LM3914 can also be cascaded to form a moving dot display of 200 or more elements. This is done by connecting pin 9 to pin 1 of the

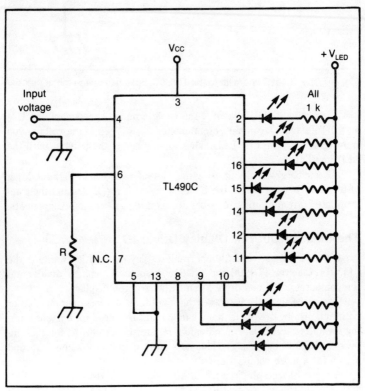

Fig. 4-9. This is a basic bargraph driver circuit using the TL490C.

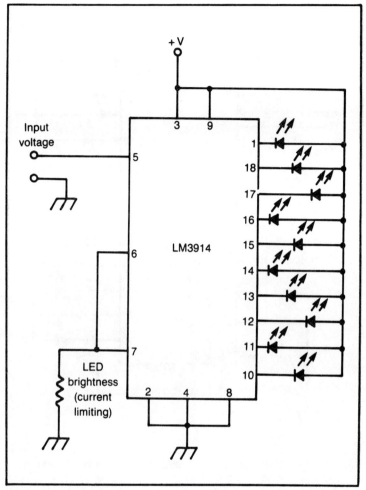

Fig. 4-10. The LM3914 is another popular bargraph driver IC.

next higher chip. This type of connection is carried through each adjacent pair of LM3914 drivers, except the last unit, on which pin 9 is connected to pin 11. Also a resistor is placed in parallel with LED 9 (from pin 11 to +Vcc) on all the chips, except for the first. This resistor should have a value of about 20 kΩ (20,000 ohms).

Closely related to the LM3914 dot/bar display driver IC is the NSM3914 dot/bar display module. This module is a small (1.99 × 0.85 inches, printed-circuit board with a LM3914 IC and a ten-element LED bargraph unit pre-mounted. This saves you the trouble of making the connections from the driver IC to each individual LED.

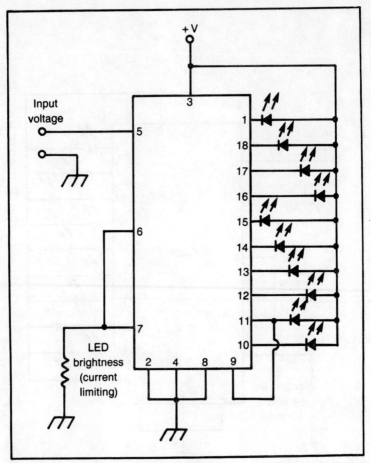

Fig. 4-11. The LM3914 can also be configured for a moving dot display.

Chapter 5

Sensor ICs

G ENERALLY, WE WANT ELECTRONIC COMPONENTS SUCH AS ICS TO BE
as insensitive as possible to any factors in their external environment.
We don't want the room temperature, or the lighting to affect the operation of the circuitry. Usually, that is. If we want to make an electronic measurement of one of these factors, we will need a device that responds to
that factor in a reliable, predictable way. In other words, in some applications we need a sensor. Several special-purpose integrated circuits have
been developed for sensor applications.

TEMPERATURE SENSORS

One of the most common types of sensor ICs is the temperature sensor, or electronic thermometer. Any solid-state component will respond to
some degree to changes in ambient temperature. Ordinarily, steps are taken
to minimize this temperature sensitivity. Component designers have done
a good job in this area.

For a temperature sensor, however, we want to take advantage of the
solid-state crystal's temperature sensitivity. The hard part is to make the
response consistent over the entire range of interest. Special-purpose thermometer ICs do very well over a wide range of temperatures. Easy, reliable temperature readings can be made efficiently, and at relatively low cost.

An IC temperature sensor converts the ambient temperature into an
electrical quantity, such as voltage, current, or frequency. This electrical
signal can then be fed to an appropriate metering circuit for a direct readout, or it can be used to drive other electronic circuits.

The Datel VFQ-1 is a temperature-to-frequency converter IC. A typical circuit built around this device is shown in Fig. 5-1. An external PNP

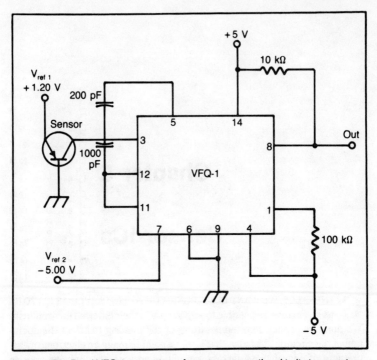

Fig. 5-1. The Datel VFQ-1 generates a frequency proportional to its temperature.

transistor (2N2907, or similar) is used as the actual temperature probe, generating a current that is proportional to the ambient temperature. The IC performs the current-to-frequency conversion operations.

The output of this circuit will have a scaling factor of 10 Hz per degree Kelvin. The Kelvin scale of temperature measurement is also known as the absolute scale. Its zero point is absolute zero, or − 459.69 degrees Fahrenheit (− 273.16 degrees centigrade). There are no negative temperatures in the Kelvin scale. A change of one degree Kelvin is equivalent to a change of one degree centigrade. By calibrating the readout circuitry so that a frequency of 2730 Hz gives a reading of 0, the circuit will read out directly in the centigrade scale.

Another temperature sensor IC is the two terminal AD590. This simple device generates a current proportional to the temperature. The scaling factor is 1 μA (0.0000001 amp) per degree Kelvin. One way of using the AD590 is illustrated in Fig. 5-2. The sensor is simply placed in series with a resistance. The potentiometer can be adjusted for calibration. Since calibration is only done at one temperature point, this method is sometimes called the one-temperature, or one-point system. Because nothing is perfect, there may be slight irregularities in response at points far removed from the calibration point.

HUMIDITY SENSOR

Electronic humidity sensors are rather complex and expensive devices, so electronic hygrometers have not been as common as, say, electronic thermometers. IC sensors may change that. With National Semiconductor's PCRC-55, a direct reading electronic hygrometer can be built for under $100.

The PCRC-55 is an electro-humidity sensor. Changes in humidity change the impedance of the surface of a chemically treated styrene copolymer plastic wafer. This device can respond very rapidly to changes in humidity. It can cover the complete range of relative humidity from 0% to 100%.

BAROMETRIC PRESSURE SENSOR

Atmospheric pressure indicates a lot about weather conditions. For example, it can be a good clue to whether or not a storm is approaching. A barometer is a device that measures atmospheric pressure. In the past we had to rely on delicate mechanical barometers. Now, solid-state barometers are becoming increasingly practical.

The LX0503A is a pressure transducer that can be used to monitor barometric pressure. It is manufactured by SenSym (1255 Reamwood Ave., Sunnyvale, CA 94809). This unit is an absolute pressure type. This means

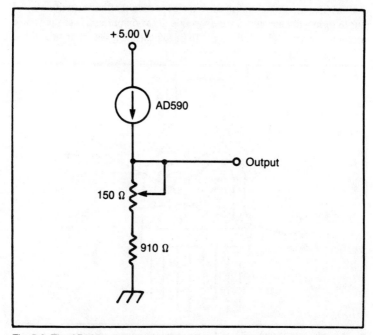

Fig. 5-2. The AD590 temperature sensor IC is placed in series with a resistance.

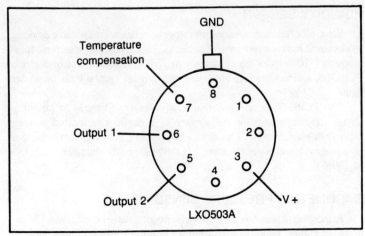

Fig. 5-3. The LX503A can be used to measure barometric pressure.

that it measures pressure relative to a vacuum. Other pressure sensors measure pressure relative to ambient pressure. Such devices are called gauges. They are not suited to barometer applications. The pinout of the 8-pin LX0503A is shown in Fig. 5-3.

Actually this device does contain a more or less mechanical pressure sensor. The sensing circuitry is on a silicon chip that has a cavity etched out to form a diaphragm. On the top side of the diaphragm is the pressure sensing circuitry. This side is said to be exposed. On the other side of the

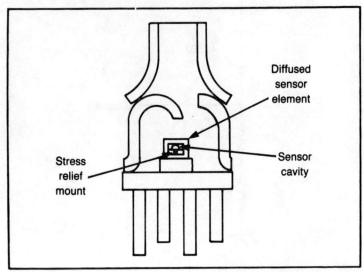

Fig. 5-4. The LX503A contains a tiny mechanical pressure-sensing diaphragm.

72

diaphragm is a small space containing a vacuum. The internal mechanical structure of this device is illustrated in Fig. 5-4.

The sensing diaphragm is deflected by the ambient pressure. Changes in ambient pressure result in changes in the amount of deflection of the diaphragm. Piezoresistive elements are used to detect the changes in deflection. The resistance changes with changes in pressure. Thanks to Ohm's law, the output voltage changes. Two outputs are provided for convenience in different applications. Output 1 (pin 6) increases with increased pressure. Output 2 (pin 5) decreases (goes negative) with increased pressure.

SMOKE DETECTOR ICs

More and more owners of homes and commercial buildings have installed smoke detectors for fire safety. It is inevitable that specialized ICs would soon appear for this application. The MD4301 is just such a device. The pinout of this CMOS IC is shown in Fig. 5-5. It is designed for use in ion-chamber type smoke detectors.

There are four output choices. The main output is at pin 8. This output can be a pure dc signal, or a combined ac/dc signal, depending on the driving requirements of the alarm indicator (horn, siren, or whatever). The

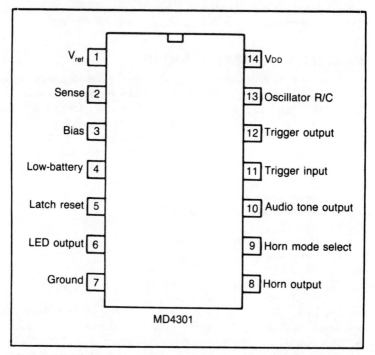

Fig. 5-5. The MD4301 is a CMOS device designed for use in ion-chamber type smoke detectors.

choice of output types is made at pin 9. Grounding this pin gives a combined ac/dc output from pin 8 when the circuit is triggered. If a logic HIGH signal is fed to pin 9, pin 8 will produce a straight dc voltage when the circuit is triggered.

The MD4301 also has an internal oscillator which can be used as an alarm indicator in some applications. The output of this oscillator is available at pin 10.

The fourth output choice is an LED at pin 6. This is especially useful in large systems with several MD4301's, or similar alarm circuits. An operator can tell at a glance which alarm is triggered by seeing which LED is on. Once triggered, the LED will remain lit until a logic HIGH signal is fed to the latch reset terminal (pin 5).

The same LED is also used as a power-on indicator. As long as power is applied, the LED will blink briefly on about once a minute. This saves the need for an extra power indicator light. Also, since the LED remains off most of the time, the power drain is significantly reduced. Of course, once the alarm is triggered the LED stops blinking, but if it is lit, the circuit is obviously getting power. It can't be confused with an actual alarm condition, which would turn the alarm device on continuously. The periodic blips of sound can't possibly be ignored or mistaken for a normal operating condition. All in all, the MD4301 is quite cleverly designed, and should be more than practical for the vast majority of smoke detector applications.

HALL-EFFECT MAGNETIC SENSOR

A voltage drop will appear across a conductor or semiconductor through which a current is flowing under the influence of a magnetic field at right angles to the direction of current flow. This is the *Hall effect*, which was discovered in 1879. It can be put to use in several electronics applications. Applications include fast switching (typical rise and fall times are 15 and 100 nanoseconds, respectively). A Hall-effect switch is bounce-free. Hall effect sensors have also been used in such applications as:

☐ Electric guitar pick-ups
☐ Telephone line current sensors
☐ Interlocks
☐ Brushless motors
☐ Sewing machines
☐ Flow meters
☐ Tire pressure gauges

The UGN-320T (Fig. 5-6) is a Hall-effect switch. In other words, it is a switch that can be controlled by an external magnetic field. Figure 5-7 is a simplified block diagram of the internal circuitry of the UGN-320T. A tiny voltage is generated by the Hall-effect sensor. This voltage is amplified and fed to a Schmitt trigger. When the voltage fed to the Schmitt

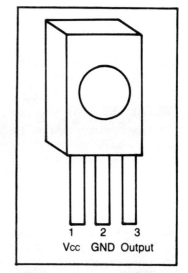

Fig. 5-6. The UGN-320T is a Hall-effect switch, which responds to changes in an external magnetic field.

1 2 3
Vcc GND Output

trigger exceeds a specific level, the circuit turns on the output transistor.

The UGN-320T uses hysteresis to prevent the output from oscillating when the controlling signal is hovering around the turn-on point. The intensity of the magnetic field must drop well below the turn-on level to turn the output transistor back off.

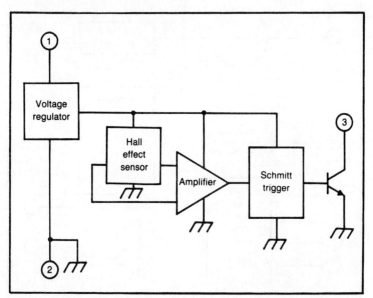

Fig. 5-7. This is a simplified block diagram of the internal structure of the UGN-320T Hall-effect switch IC.

75

The power requirements of this IC are very flexible, thanks to the on chip voltage regulator. The supply voltage can be anything from +4.5 volts to +20 volts. The UGN-320T is designed for low power consumption. Typically, the current draw is only about 12 mA (0.012 amp), assuming the supply voltage is +12 volts.

METAL SENSING ICs

Another fairly common electronic sensing application is metal detection—an electronic circuit that will give an indication when it is brought near an object made of conductive metal. The metal detectors used in airports are one common way of putting this type of circuit to use. Not surprisingly, ICs for metal detectors have been developed. Where there is an application, sooner or later there is an IC to do the job.

A block diagram of the internal circuitry of the CS191 metal detector IC is shown in Fig. 5-8. A slightly similar device of the same type is the CS209. Figure 5-9 illustrates the internal circuitry of this chip. Both of these chips are manufactured by Cherry Semiconductor Corp. (2000 South County Trail, E. Greenwich, RI 02818).

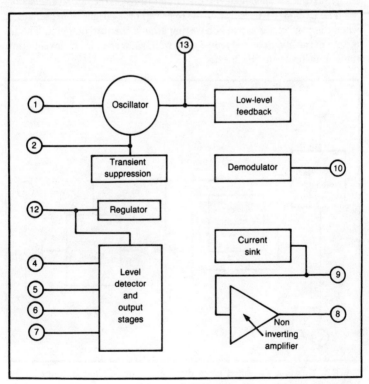

Fig. 5-8. The CS191 is a special-purpose IC designed for use in metal detectors.

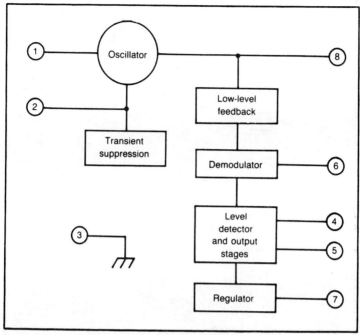

Fig. 5-9. Another metal detector IC is the CS209.

AUTOMATIC BRIGHTNESS CONTROL

The Optron OPL100 is a rather unique device that senses the level of light shining on it, and produces an output that is proportional to the detected brightness. It can adjust the brightness of an external light source (such as an incandescent bulb, or an LED) in response the ambient light level. Its control range runs from 0% to 100%. The OPL100 can control many different types of devices, including:

☐ Incandescent lights
☐ Indoor or outdoor lighting
☐ Industrial lighting levels
☐ Instrument displays in automobiles and airplanes
☐ LEDs
☐ Security lighting
☐ Signs and advertising displays
☐ Vacuum fluorescent displays

The pinout diagram of the OPL100 is shown in Fig. 5-10. The internal circuitry of this device is illustrated in Fig. 5-11. The OPL100 can be used to create a consistent level of light indoors, despite changes in the light level from outdoors (through windows, or whatever).

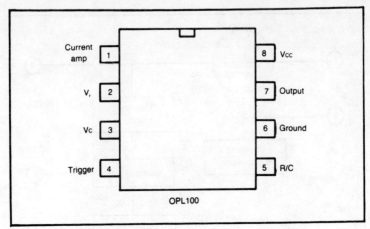

Fig. 5-10. The OPL100 is an automatic brightness control IC.

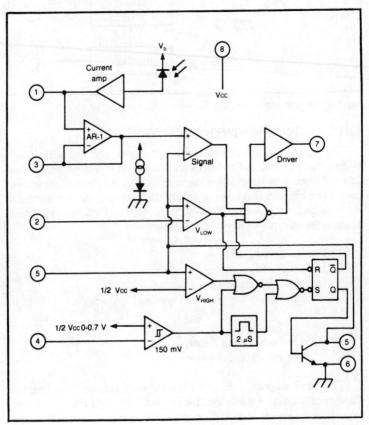

Fig. 5-11. This is a block diagram of the internal structure of the OPL100.

The "secret" of this device is pulse-width modulation. By varying the duty cycle (ratio of ON time to total cycle time) of a rectangle wave, the average output voltage can be made proportional to the ambient light level. The pulse width is determined by the analog voltage from the built-in light sensor and the instantaneous level of an internally generated sawtooth wave. The comparator switches on when the sensor voltage is greater than the instantaneous sawtooth voltage. When the sensor signal is less than the sawtooth, the comparator switches off. Figure 5-12 illustrates how this method can produce rectangle waves with pulse widths that are proportional to the sensor voltage. The pulse-width modulated signal is the out-

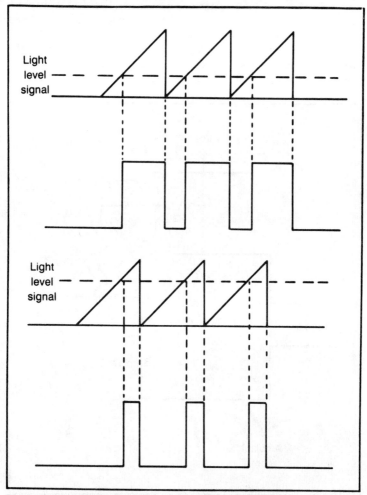

Fig. 5-12. The OPL100 generates a rectangle wave with a pulse width proportional to the amount of light falling on the sensor element.

put from the chip at pin 7. The ambient light level is detected by a tiny built-in photodiode. This sensor is only 2500 square mil (0.0025 square inch).

The circuit shown in Fig. 5-13 shows how the OPL100 can be used to automatically control the brightness of an LED, in response to the ambient lighting. With a 12-volt power supply, the OPL100 will feed a maximum of 32 mA to the LED. Potentiometer R1 is a light sensitivity control. The resistance value needed here will depend on the desired brightness of the LED, and the average ambient lighting level. Typical values range from 25 kΩ to 200 kΩ.

Another typical circuit built around the OPL100 is shown in Fig. 5-14. This circuit controls the brightness of a vacuum fluorescent display. R1 is again a sensitivity control. R2 prevents the display from being turned off when the sensor is in total darkness.

The sensor is positioned so that it is "looking" in the direction of the display. Once R1 is adjusted for the desired brightness level, the OPL100 will monitor the ambient lighting level, and adapt the brightness of the dis-

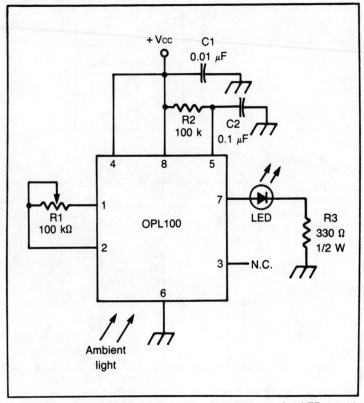

Fig. 5-13. This circuit automatically controls the brightness of an LED according to the overall illumination.

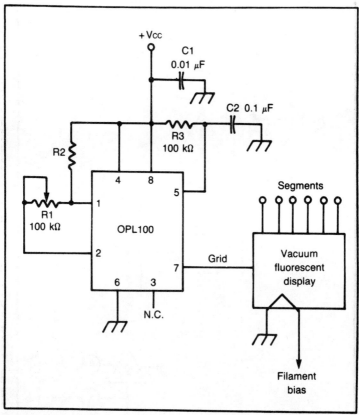

Fig. 5-14. The OPL100 can be also used to control the brightness of a vacuum fluorescent display.

play accordingly. Therefore, the display's brightness will appear to be constant.

MOTION DETECTOR

Another specialized IC with a built-in light sensor is the D1072. It is designed as a motion detector, primarily for use in security alarm systems. A built-in lens in this IC allows the light sensor to "see" a two foot circle at a distance of eight feet. The monitored area is conical, as illustrated in Fig. 5-15. If anything moves within this protected area, the amount of light reaching the sensor will change slightly. The D1072 can respond to changes as small as 5%. Moreover, the change can be in either direction. It will respond if the light level either increases or decreases.

No special external light source is required for operation. The D1072 is designed to function in ambient light. Its usable range covers a 1000:1

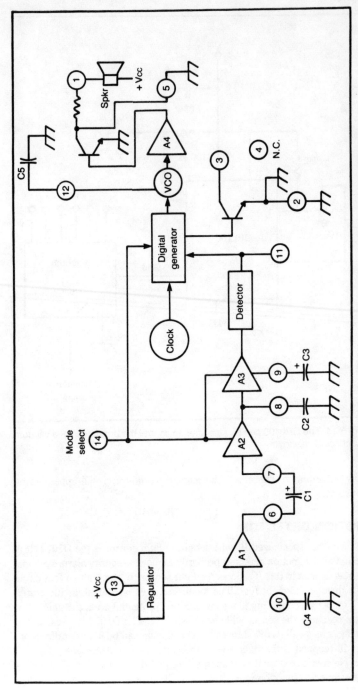

Fig. 5-15. The D1072 detects motion by watching for changes in the amount of light striking its sensor element.

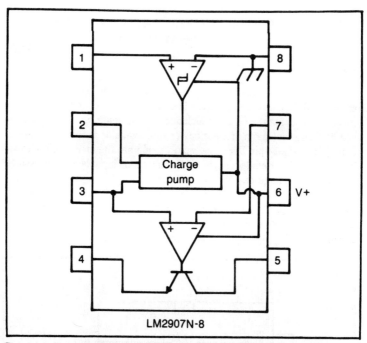

Fig. 5-16. The LM2907N-8 is a tachometer IC in an 8-pin DIP housing.

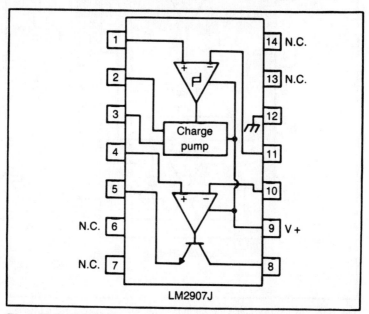

Fig. 5-17. The LM2907J is a 14-pin version.

83

light-level range. It can function in lighting as low as 0.1 candlepower (very dark) to as high as 100 candlepower (very bright).

TACHOMETER ICs

National Semiconductor makes two tachometer/speed switch ICs. They are the LM2907 and the LM2917. The primary difference between these two devices is that the LM2917 includes internal zener regulation, and the LM2907 does not. Both of these devices are available in two versions—an 8-pin DIP and a 14-pin DIP. The 8-pin version has the inverting reference input internally grounded. In the 14-pin units, this input is user accessible. The 14-pin package has four unused pin terminals.

The 8-pin LM2907N-8 is shown in Fig. 5-16. Figure 5-17 shows the 14-pin LM2907J. The 8-pin LM2917N-8 is shown in Fig. 5-18, and the 14-pin LM2917J is shown in Fig. 5-19. These devices use a frequency-to-voltage converter to change the ac signal from an external mechanical sensor to

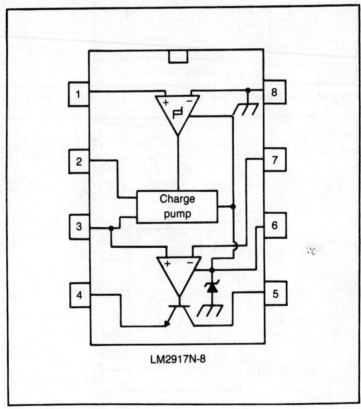

LM2917N-8

Fig. 5-18. The LM2917N-8 is similar to the LM2907N-8, except that it also features internal zener regulation.

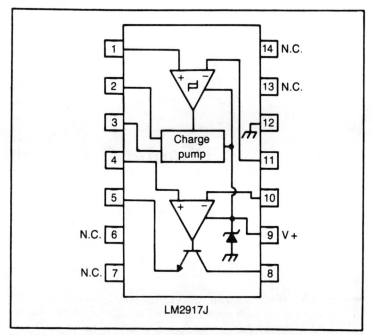

Fig. 5-19. The LM2917J is contained in a 14-pin DIP housing.

a constant-width output pulse. The pulse width is selected so that the average current level of the output signal is proportional to the input frequency. Hysteresis is used to reduce sensitivity to noise, which is likely to be present in most tachometer applications.

Chapter 6

Amplifiers

P ROBABLY THE MOST BASIC AND FREQUENTLY ENCOUNTERED ELEC-
tronics application is amplification. An amplifier is a device that increases the amplitude of an electrical signal. In this chapter we will be concentrating primarily on audio amplifiers, although a few rf (radio frequency)
amplifier ICs will also be presented.

A great many amplifier ICs have been developed. Some are extremely
impressive in how much power they can pack into such a small space. For
the most efficient performance, IC amplifiers should be used with adequate
heatsinking. In most cases, the better the heatsinking, the greater the maximum output power will be.

THE LM380 AUDIO AMPLIFIER IC

An audio amplifier IC that has been around for a few years, and still
enjoys considerable popularity is the LM380. It is widely available in two
types of packaging. The pinout for the 8-pin DIP version is shown in Fig.
6-1, while the 14-pin DIP version is illustrated in Fig. 6-2. On both chips,
only six of the pins are active. The remaining pins are shorted to ground,
and provide some internal heatsinking. The 14-pin version has more "heatsink" pins than the 8-pin version and it can handle greater amounts of power
without overheating.

Without an external heatsink, the LM380 can dissipate up to about
1.25 watts at room temperature. Higher output levels can be achieved with
external heatsinking. For example, if the six "heatsink" pins of a 14-pin
LM380 are soldered to a six square inch copper foil pad on a pc board (two-
ounce foil), the IC can produce up to 3.7 watts at room temperature.

This chip features an internal automatic thermal shutdown circuit which

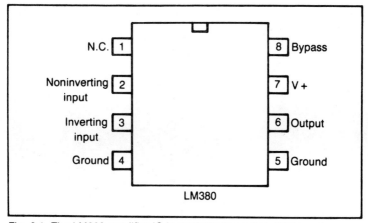

Fig. 6-1. The LM380 amplifier IC is available in an 8-pin DIP housing.

will turn the amplifier off if excessive current causes the IC to start overheating. The internal circuitry is made up of twelve transistors, but a simplified version of the circuitry is shown in Fig. 6-3.

Gain is internally fixed at 50 (34 dB). The output automatically centers

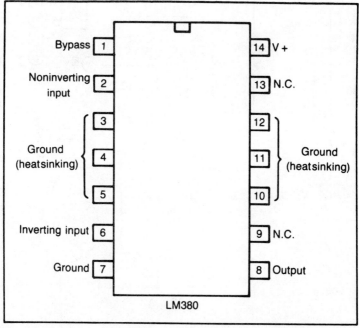

Fig. 6-2. Even the 14-pin DIP version of the LM380 only has six active pins. The extra pins are used for heatsinking.

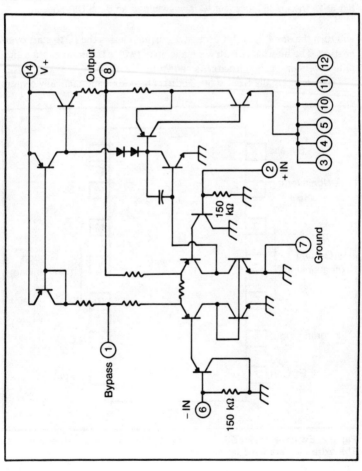

Fig. 6-3. This is a simplified diagram of the internal circuitry in the LM380 amplifier IC.

itself at one half of the supply voltage, eliminating offset problems. If a symmetrical dual-polarity power supply is used, the output will be centered around ground potential (0 volts) with no dc component.

The input stage of the LM380 is rather unique. The input signal may be either ground referenced, or ac coupled, depending on the requirements of the specific application.

The inputs are internally biased with a 150 kΩ resistance to ground. Transducers which are referenced to ground (no dc component) may be directly coupled to either the inverting or the noninverting input. There are several possibilities for handling the unused input terminal. They are:

☐ Leave it floating
☐ Short it directly to ground
☐ Reference it to ground through a resistor or capacitor

In most applications in which the noninverting input is used, the inverting input will be left floating. This may make the board layout critical. Stray capacitances may lead to positive feedback and undesired oscillations.

The LM380 is designed for use with a minimum of external components. The most basic form of a LM380-based amplifier circuit is shown in Fig. 6-4. Clearly, it would be hard for things to be much simpler than this. The only external component is the output decoupling capacitor.

Of course, in many actual applications additional external components will be required. For instance, if the chip is located more than two or three inches from the power supply's filter capacitor, a decoupling capacitor should be mounted between the V + terminal of the LM380 and ground.

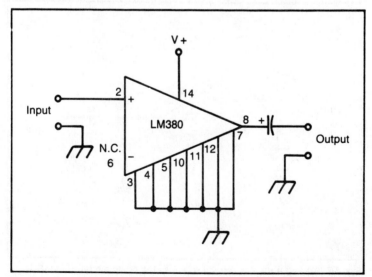

Fig. 6-4. This is the most basic circuit built around the LM380 amplifier IC.

This decoupling capacitor should be mounted as physically close to the IC as possible. A typical value would be 0.1 μF.

If the LM380 is to be used in a high-frequency application (several megahertz or more) it may break into oscillations. Adding the extra resistor and capacitor shown in Fig. 6-5 between the output and ground will help suppress such oscillations. Generally, the resistor will be quite small (typically 2.7 ohms). The capacitor's value should be 0.1 μF.

Since these oscillations occur at 5 to 10 MHz, they won't be of much significance in audio applications, but if the LM380 is being used in an rf sensitive environment, they may pose a problem unless suppressed. (Note that in Fig. 6-5, and in future diagrams, the heatsinking pins are not shown, for simplicity.)

Figure 6-6 shows a practical audio amplifier circuit using the LM380. The input is provided by an inexpensive low-impedance microphone. An impedance matching transformer should be used for low-impedance sources. The 1 MΩ potentiometer serves as a volume control. An 8-ohm speaker can be driven directly by the LM380 audio amplifier IC. The fixed gain of this device can be increased with positive feedback.

The LM380 is often used in inexpensive tape recorders and phonographs. Figure 6-7 shows a simple phonograph amplifier, driven by a ceramic cartridge. Potentiometer R1 is a simple voltage-divider volume control. Potentiometer R3 is a basic tone control, making the high frequency roll-off characteristics of the circuit manually adjustable.

Most phonograph applications require frequency response shaping to provide the standard RIAA equalization characteristic. All commercial

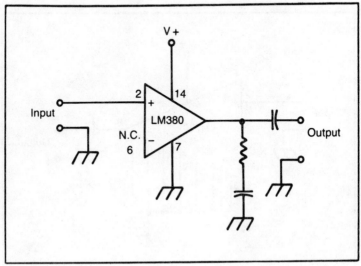

Fig. 6-5. Adding a resistor and capacitor to the basic circuit of Fig. 6-4 can help minimize oscillation problems.

90

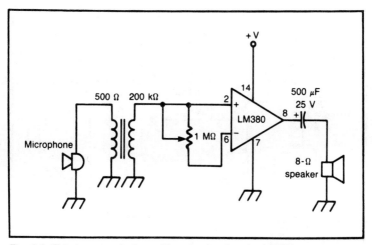

Fig. 6-6. This is a practical amplifier circuit using the LM380.

records are equalized according to the RIAA standards, so if the complementary response shaping isn't included in playback the sound won't be very good. Figure 6-8 shows a LM380 phonograph amplifier with RIAA equal-

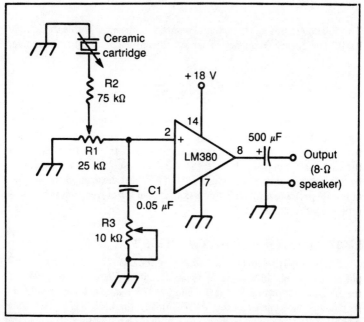

Fig. 6-7. The LM380 is ideal for use in inexpensive ceramic cartridge phonographs.

91

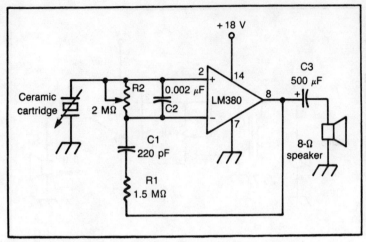

Fig. 6-8. This phonograph amplifier includes RIAA equalization.

ization. The mid-band gain can be defined with this formula:

$$G_{mid} = \frac{R1 + 150,000}{150,000}$$

The 150,000 represents the internal resistance presented within the LM380 itself. The corner frequency is set via capacitor C1, using this formula:

$$F_c = \frac{1}{2\pi C1R1}$$

A pair of LM380 amplifiers can be put into a bridge configuration, as shown in Fig. 6-9 to achieve more output power than could be obtained from a single amplifier. This circuit provides twice the voltage gain across the load for a given supply voltage. This increases the power capability by a factor of four over a single LM380. (The heat dissipation capabilities of the IC package may limit the maximum output power below the theoretical quadruple level.)

LM377/378/379 POWER AMPLIFIERS

Fairly closely related to the LM380 are the LM377/378/379 power amplifier ICs. The LM377 is a dual 2-watt amplifier. The pinout diagram for the LM377 is shown in Fig. 6-10. The LM378 is a dual 4-watt amplifier. Its pinout is the same as for the LM377. Finally, the LM379 is a dual 6-watt amplifier, and its pinout is shown in Fig. 6-11. Each of these ICs contain two complete amplifier stages and they are ideally suited for stereophonic

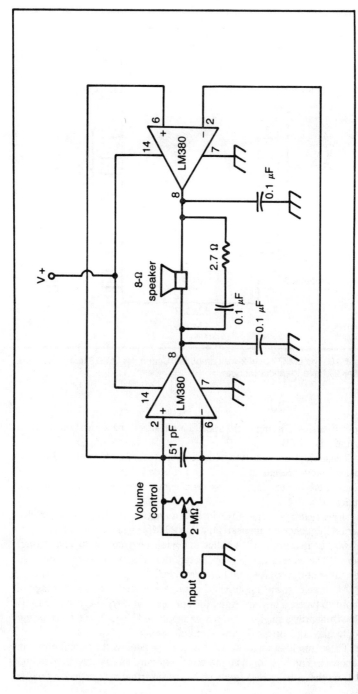

Fig. 6-9. A pair of LM380s in a bridge configuration can produce more power.

93

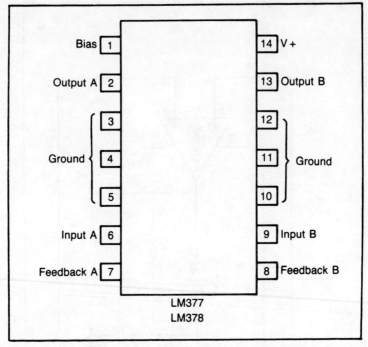

Fig. 6-10. The LM377 dual 2-watt amplifier IC, and the LM378 dual 4-watt amplifier IC have identical pinouts.

applications. According to the specification sheet, the separation between channels is 75 dB.

These chips are designed to drive 8- or 16-ohm loads directly. Because most speakers present an 8- or 16-ohm impedance, no output transformers are required. In fact, very few external components are required for most applications.

Other features of these ICs include a typical open-loop gain of 90 dB, internal frequency compensation, a 5 to 20 MHz gain-bandwidth product, and fast turn-on/turn-off, without annoying (and potentially damaging) "pops." The outputs are fully protected. Both output current-limiting and thermal shut-down protection are provided.

All three of these amplifier ICs can be operated with a wide range of supply voltages. They are rated for operation at anything from 10 to 35 volts. Ordinarily a single-ended power supply will be used, but a dual power supply may also be used without problems.

These amplifiers may be used in either a noninverting configuration, as shown in Fig. 6-12, or in an inverting configuration, as illustrated in Fig. 6-13. As these diagrams clearly show the inverting version has a smaller external parts count, so it will tend to be less expensive when driven by a low impedance source.

Figure 6-14 shows how one of these ICs can be used to build a stereo amplifier in the noninverting mode. An inverting stereo amplifier is illustrated in Fig. 6-15. The feedback resistors (R_f) shouldn't be any larger than about 1 MΩ, or severe distortion may result.

These ICs will place very little loading on their signal sources. The input impedance is in the neighborhood of 3 MΩ (3,000,000 ohms).

Adding output transistors can increase the power handling capability of these amplifiers. Figure 6-16 shows the LM378, which normally produces 4 watts, in an amplifier circuit that can put out 10 to 12 watts. Not only is this circuit quite simple, it also exhibits even less crossover distortion than the LM378 alone.

THE μA783 AUDIO AMPLIFIER IC

It is really quite remarkable how much power IC designers are able

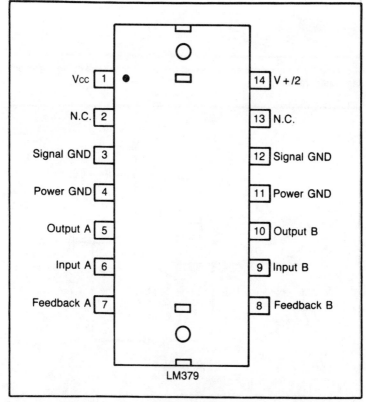

Fig. 6-11. The LM379 is a dual 6-watt amplifier IC.

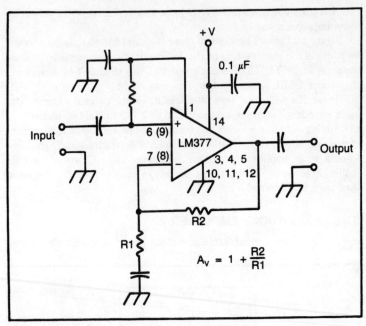

Fig. 6-12. There is no phase shift when the amplifier is used in the noninverting configuration.

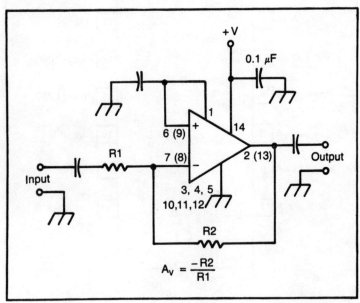

Fig. 6-13. In the inverting configuration, the signal is phase shifted 180 degrees.

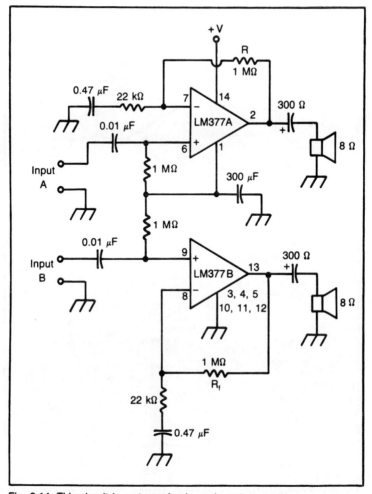

Fig. 6-14. This circuit is a stereophonic noninverting amplifier.

to pack into a tiny silicon chip. The μA783 is contained in a small 12-pin package that is less than an inch long and a quarter inch wide. It is a full 9-watt amplifier. In most applications, the largest thing in the circuit would be the heatsink to prevent the IC from overheating. The μA783 audio amplifier is designed to drive 8-ohm or 16-ohm speakers. It is truly a high fidelity device, with less than 0.3% of harmonic distortion.

THE LM381 LOW-NOISE DUAL PREAMPLIFIER

The signal from a magnetic tape head, or a magnetic type phonograph cartridge is very small. The noise generated by the amplifier itself may

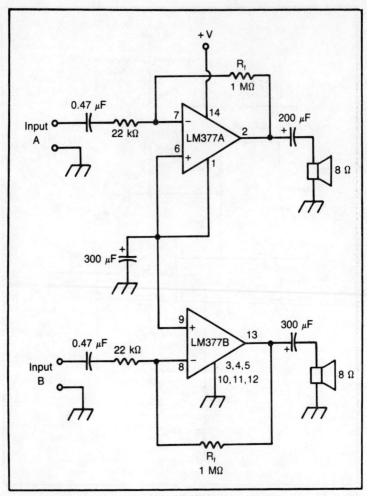

Fig. 6-15. This is a stereophonic inverting amplifier circuit.

become critical. Too much amplifier noise can completely overpower the desired signal.

Low-level signal sources usually require some sort of preamplifier. A preamplifier is simply an amplifier stage designed for the best possible signal to noise ratio. The preamplifier boosts the weak signal up to a level at which it can be handled by the power amplifier.

The LM381 contains two high fidelity low-noise preamplifier stages in a single 14-pin package. The pinout diagram for this device is shown in Fig. 6-17. The LM381 generates a minimum of noise. A typical value for the noise is 0.5 μV (0.0000005 volt) rms.

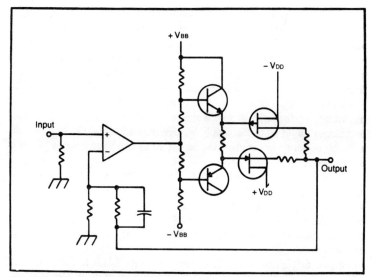

Fig. 6-16. The output power of an amplifier IC can be increased by adding external output transistors.

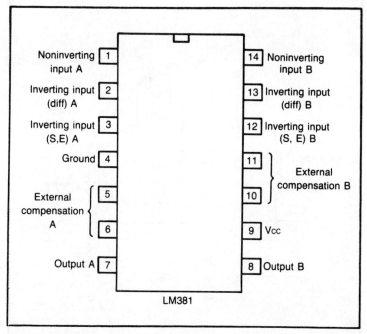

Fig. 6-17. The LM381 contains two high fidelity low-noise preamplifier stages in a single 14-pin package.

99

Two completely independent preamplifier stages are contained in the LM381. An internal power supply decoupler/regulator results in 120 dB supply rejection. In other words, ripple in the supply voltage will have a minimal effect on the operation of the circuit. A wide range of power supply voltages (9 to 40 volts) can be used. The two channels are well isolated from each other. The channel separation specification for this chip is 60 dB.

The LM381 provides a fairly hefty amount of gain—112 dB. The output voltage can swing over a wide range. The peak-to-peak voltage of the output can swing up to the supply voltage minus 2 volts. For example, if a 15-volt power supply is used, the output signal can be as large as 13 volts peak-to-peak.

This chip is designed primarily for high fidelity audio applications, but it is also suitable for wide-band instrumentation applications. The small signal bandwidth is 15 MHz, which carries the LM381's capabilities well beyond the audio range. Like most modern amplifier ICs, the LM381 features output short circuit protection. This IC is also internally frequency compensated.

Figure 6-18 shows a typical application for the LM381. This is an amplifier for a magnetic tape head. It is designed for more or less flat frequency response. The mid-band gain is determined by the resistor values.

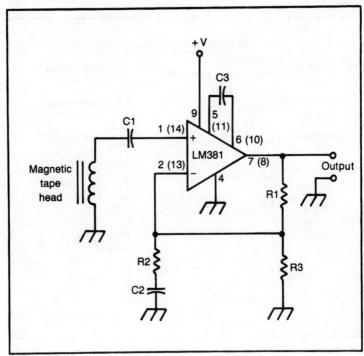

Fig. 6-18. The LM381 is a fine choice for a preamplifier in a tape recorder.

100

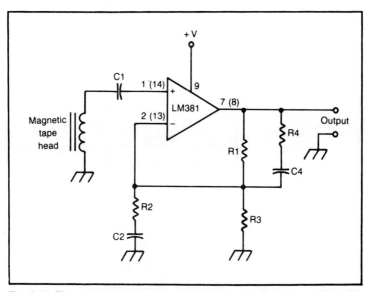

Fig. 6-19. This tape playback preamplifier includes NAB frequency compensation.

The gain formula is:

$$A_v = \frac{R1\ R2}{R2}$$

The low frequency 3 dB corner frequency is defined as:

$$F_c = \frac{1}{2\ \pi\ C2R2}$$

This formula is valid only if the capacitive reactance of C2 (X_{c2}) is equal to the resistance of R2.

Most modern tape recorders/players use standard NAB equalization. Figure 6-19 shows the LM381 in a NAB compensated magnetic tape playback amplifier.

Most phonograph and tape player amplifiers include tone (bass and treble) controls to allow the user to customize the sound to his personal taste and/or the specific requirements of the environment. Generally, because of the insertion loss of the tone control, it is usually placed between two preamplifier stages. Thanks to the high gain capabilities, and large output capability of the LM381, only a single preamplifier stage is required, even with passive tone controls. Figure 6-20 shows a simple bass control network for use with the LM381. Figure 6-21 shows a similar treble control network.

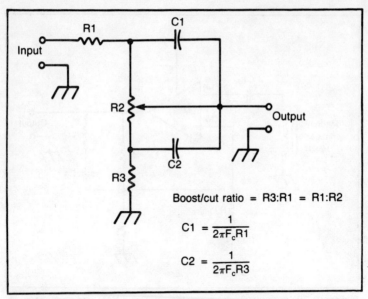

Boost/cut ratio = R3:R1 = R1:R2

$$C1 = \frac{1}{2\pi F_c R1}$$

$$C2 = \frac{1}{2\pi F_c R3}$$

Fig. 6-20. This is a simple bass control network for use with the LM381.

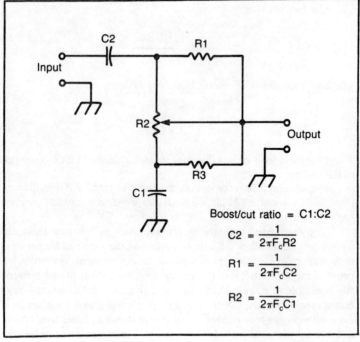

Boost/cut ratio = C1:C2

$$C2 = \frac{1}{2\pi F_c R2}$$

$$R1 = \frac{1}{2\pi F_c C2}$$

$$R2 = \frac{1}{2\pi F_c C1}$$

Fig. 6-21. This simple treble control network can be used with the LM381.

102

THE ICL8063 POWER TRANSISTOR DRIVER/AMPLIFIER

Most IC amplifiers are designed for relatively low-power applications. For a high-power output, external output transistors are needed. The ICL8063 (shown in Fig. 6-22) is a monolithic amplifier intended to drive power transistors for high-wattage outputs. It is intended to drive complementary-symmetry outputs in audio amplifiers. The ICL8063 can also be used as a driver for servo and stepping motors, and rotary or linear actuators. This device takes signals in the ± 11 volts range, and amplifies them into the ± 30 volt range at 100 mA to drive power transistors.

The ICL8063 uses a ± 30 volt power supply. It includes on-chip ± 13-volt regulators. This chip is compatible with most op amps, preamps, companders, and similar IC devices.

Figure 6-23 shows the ICL8063 in a 50-watt rms amplifier circuit, with an 8-ohm load. Distortion is less than 0.1% for frequencies below about 100 Hz. It only goes up to 1% at 20 kHz (20,000 Hz).

The two 0.4 Ω resistors are for current limiting. The 1000 pF capacitors help maintain good stability down to unity gain. Of course the transistors and the IC should be adequately heatsinked.

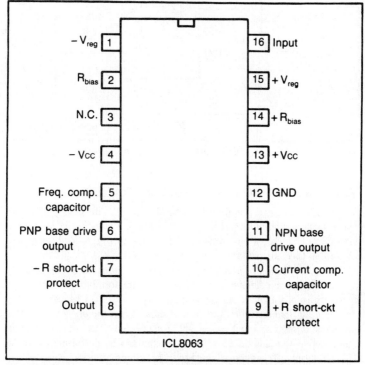

Fig. 6-22. The ICL8063 is designed to drive high power transistors for a large wattage output.

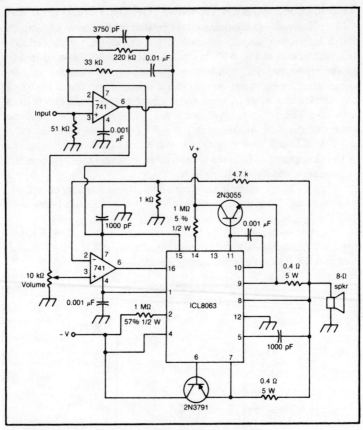

Fig. 6-23. This circuit will put out 50 watts to an 8-ohm load.

THE HA-2400 PROGRAMMABLE AMPLIFIER

A rather unusual amplifier IC is the HA-2400. This device is known variously as a *programmable amplifier* (PRAM), or a *four-channel operational amplifier*. This chip's pinout diagram is shown in Fig. 6-24. Figure 6-25 illustrates the internal structure of this device.

Basically, there are four op-amp input stages, which are selectable via a digitally controlled electronic switch. One of the four input stages drives a fifth op amp, which serves as an output stage. Table 6-1 lists the various digital input combinations, and their results. When the enable pin (14) is HIGH, one of the four input stages will be connected to the input of the output stage. Any standard op-amp application can be performed with the HA-2400, with the added advantage of programmability.

The output stage is internally wired as a unity gain voltage follower, so feedback components can be connected from the chip output (pin 10) to

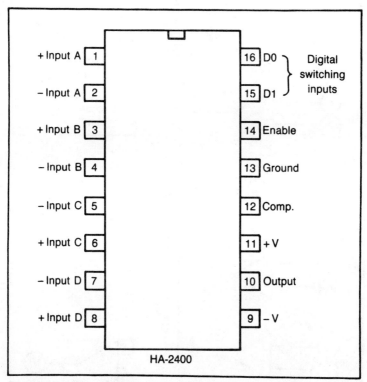

Fig. 6-24. The HA-2400 Programmable Amplifier IC is also called a four-channel op amp.

the appropriate input, and the device will function as a regular op amp, depending on which input stage is selected.

If the HA-2400 is being used with a gain less than 10, frequency compensation should be used to ensure closed-loop stability. This is done by

Table 6-1. Different Digital Inputs Are Used to Select the Various Input Stages of the HA-2400 PRAM.

Digital Inputs			Channel			
D0	D1	Enable	A	B	C	D
X	X	0	Off	Off	Off	Off
0	0	1	Off	Off	On	Off
0	1	1	On	Off	Off	Off
1	0	1	Off	Off	Off	On
1	1	1	Off	On	Off	Off

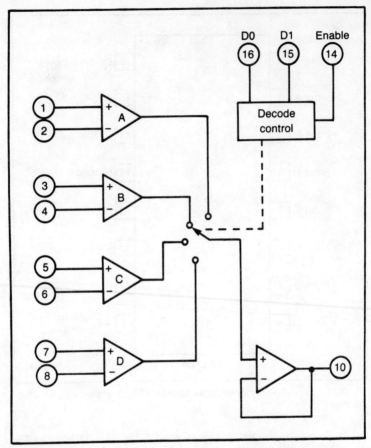

Fig. 6-25. The HA-2400 contains four digitally selectable input stages.

connecting a small capacitor (2 to 15 pF) from pin 12 to ac ground (the V + supply is recommended by the manufacturer). Each of the four input stages could be wired for different op-amp applications, allowing the function to be digitally selectable.

The circuit designer must bear in mind that the unselected input stages may still constitute a load at the amplifier output and the signal input. The analog input terminals of an OFF channel draw the same bias current as an ON channel. The input signal limitations must be observed even when the channel is OFF.

When pin 14 (ENABLE) is made LOW (grounded) all four input channels are OFF. The output voltage (at pin 10) will tend to slowly drift towards the negative supply voltage (– V). If the specific application requires a zero-volt output, wire one of the input channels as a voltage follower with the noninverting input grounded (input of 0 volts). Select this "dummy" channel instead of deactivating the ENABLE input of the chip.

It is usually not possible to wire the outputs of two or more HA-2400's together. This is because the output impedance remains low, even when the inputs are disabled. One way around this problem is to use the compensation pin (12) as the output. The voltage at this pin will be about 0.7 volt higher than at the output pin (10), but the output impedance here is very high. Consequently, two or more compensation outputs may be wired together.

The HA-2400 is a very handy and economical device. The programmability results in a tremendous amount of versatility in circuit design. Even in applications where only a single channel is to be switched on and off, it will often be less expensive to use a HA-2400 than a separate op amp and analog switch.

Any unused digital inputs must be shorted to ground (for a permanent LOW state), or to +5.0 volt (for a permanent HIGH). If they are left floating, they will behave as if held HIGH. The digital inputs are DTL and TTL compatible.

Figure 6-26 shows a typical application for the HA-2400 PRAM. This

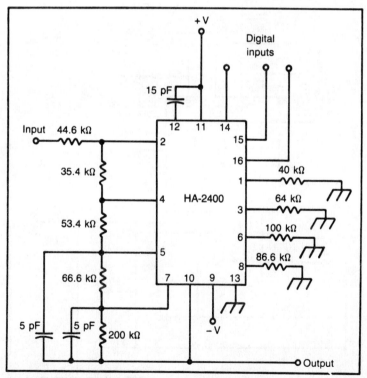

Fig. 6-26. This inverting amplifier circuit with programmable gain is a typical application for the HA-2400 PRAM.

107

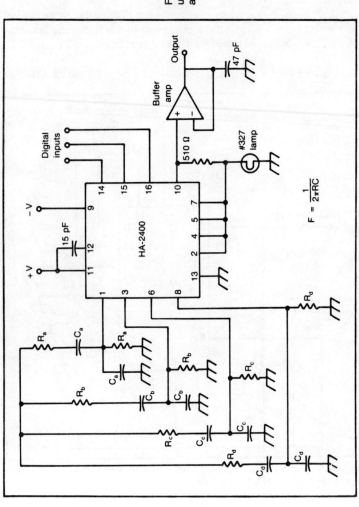

Fig. 6-27. The HA-2400 can also be used as a sine-wave oscillator with a programmable output frequency.

$$F = \frac{1}{2\pi RC}$$

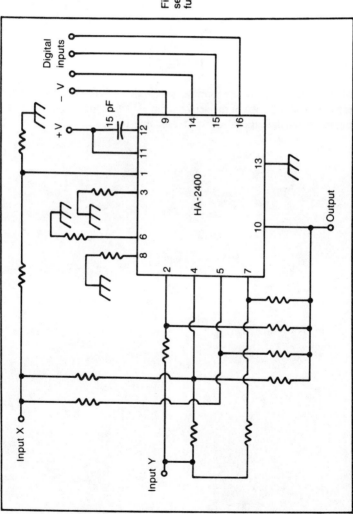

Fig. 6-28. This circuit can perform several different adder/subtractor functions.

is an inverting amplifier with programmable gain. Depending on the states of the digital inputs, the input signal will be given a gain of 0, −1, −2, −4, or −8. The same thing could also be done with one input resistor and one feedback resistor per channel, but this circuit has a lower parts count— only five resistors instead of eight.

Figure 6-27 shows the HA-2400 being used as a sine-wave oscillator with a programmable output frequency.

The circuit shown in Fig. 6-28 is a multifunction circuit. It can perform various adder/subtractor functions, depending on the states of the digital inputs. The four possible functions that may appear at the output of this circuit are:

$$-G1X$$
$$-G2Y$$
$$-(G3X + G4Y)$$
$$-G5X - G6Y$$

where G stands for the appropriate gain, which will be dependent on the resistance values used in the active channel. These are just a few of the many, many applications for this fascinating IC.

Chapter 7

Signal Generators

I N MANY ELECTRONICS APPLICATIONS, THE SIGNAL(S) BEING OPERATED
on are from an external source. A transducer or sensor (see Chapter
5) convert some quantity in the real world (temperature, light, magnetic
flux, etc.) into an electrical signal (voltage, current, etc.). In other applications we will want to work with an electronically generated signal. In this
chapter we will consider signal generator ICs.

In the broadest sense, a signal generator is an oscillator. It is a circuit
that produces an ac signal of a specific waveshape. Back in Chapter 3 we
covered rectangle wave generators (astable multivibrators). In this chapter we will be looking at more sophisticated signal generators.

FUNCTION GENERATORS

A function generator is a circuit that is capable of generating two or
more of the basic waveshapes. The simplest waveshape is the sine wave,
which is illustrated in Fig. 7-1. This is the only truly pure signal. It consists of just a single frequency component. There is no harmonic content.

The triangle wave, shown in Fig. 7-2, can sometimes be substituted
for a sine wave, since its harmonic content is relatively weak, and can virtually be eliminated by passing the signal through a low-pass filter.

Rectangle waves (Fig. 7-3) were covered in Chapter 3 in the discussion
of astable multivibrators. The signal switches quickly back and forth between two discrete levels. A square wave (Fig. 7-4) is a special case of the
rectangle wave in which the LOW level time is equal to the HIGH level time.
Rectangle waves are relatively rich in harmonic content.

The last basic waveshape is the sawtooth wave. It is sometimes called

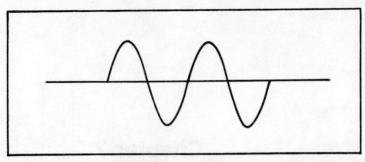

Fig. 7-1. The sine wave is the only pure waveshape with only a single frequency component.

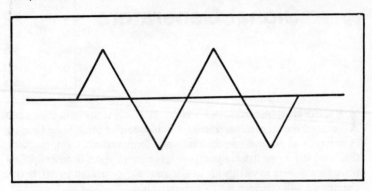

Fig. 7-2. The triangle wave has a weak set of harmonics.

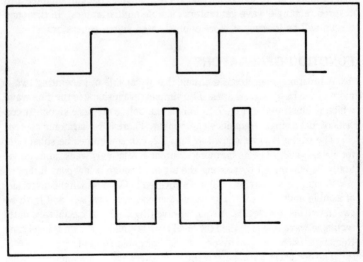

Fig. 7-3. Rectangle waves can have different duty cycles or pulse widths.

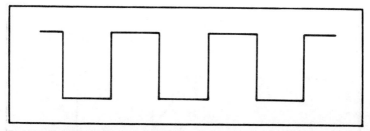

Fig. 7-4. A square wave is a special type of rectangle wave, which has a duty cycle of 1/2.

a ramp wave. This signal has a very rich harmonic content. Sawtooth waves appear in two forms. An ascending sawtooth wave is shown in Fig. 7-5, and Fig. 7-6 shows a descending sawtooth wave.

A number of function generator ICs for producing these (and other) waveforms have been developed over the years. We will look at just two of them.

The 8038 Function Generator IC

Quite a bit of versatility is packed into the 14-pin package of the 8038 function generator IC. The pinout diagram for this powerful chip is shown in Fig. 7-7. A block diagram of the 8038's internal circuitry is shown in Fig. 7-8.

The 8038 can generate sine waves, triangle waves, sawtooth waves, and rectangle waves with almost any duty cycle. The output frequency can range from 0.001 Hz to over 1.0 MHz (1,000,000 Hz) with a typical distortion rating of 1.0%, or less. External voltage control can be used to adjust the output frequency. The nominal output frequency is set via two external timing resistors and a single timing capacitor. A wide frequency range can be achieved with a single capacitance value.

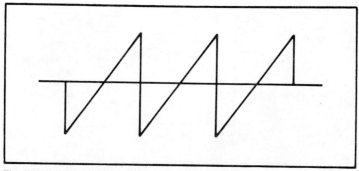

Fig. 7-5. An ascending sawtooth wave is a waveshape that is very rich in harmonics.

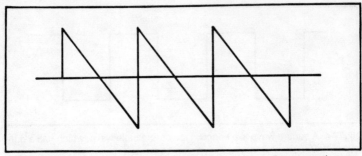

Fig. 7-6. A descending sawtooth wave is similar to an ascending sawtooth wave.

The overall specifications for this chip are quite impressive. The linearity is better than 0.1%. The 8038 is also extremely drift free. According to the manufacturer, the frequency drift is no more than 50 ppm (parts per million) per degree centigrade. This may also be written as 0.005% per degree centigrade. That much frequency drift can be reasonably ignored in the vast majority of applications.

The power-supply requirements are quite flexible. A dual-polarity power

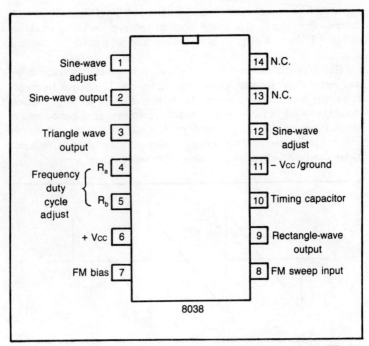

Fig. 7-7. The 8038 function generator IC can produce several different waveshapes.

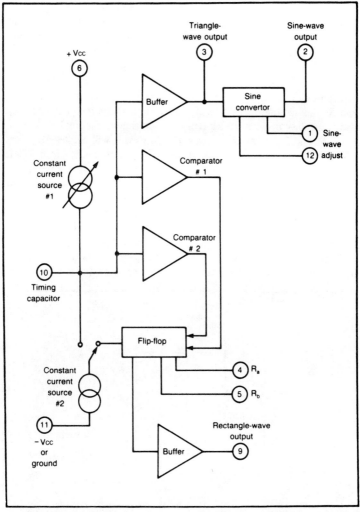

Fig. 7-8. This is a block diagram of the 8038 function diagram IC.

supply from ± 5 volts to ± 15 volts or a single-ended power supply from + 10 volts to + 30 volts may be used to operate the 8038.

If a single-ended supply is used, the output signal will be symmetrical around one-half the supply voltage. If for example, a + 18 volt supply is used, the output will be symmetrical around + 9 volts. The rectangle wave output (pin 9) switches back and forth between ground and Vcc. An external pull-up/load resistor is used on this output. The resistor does not necessarily have to reference the rectangle wave output to Vcc. For instance, if the pull-up resistor is connected to a separate + 5 volt source, the rectangle wave output will be TTL compatible (switching between 0 and +5 volts), regardless of the supply voltage. If a dual-polarity power supply is used to drive the 8038, all of the waveform outputs will be symmetrical around ground potential (0 volts).

The basic arrangement of the timing (frequency determining) components is illustrated in Fig. 7-9. The output frequency is dependent on the values of C, R_a, and R_b. In addition, the ratio between the values of R_a and R_b defines the duty cycle of the output waveform. For a rectangle wave, the duty cycle is the ratio of the HIGH time to the total cycle time.

If R_a and R_b have equal values, the duty cycle will be 50%. By varying the ratio of these resistance values, duty cycles ranging from 2% to 98% can be achieved. The duty cycle also affects the sine wave (pin 2) and triangle wave (pin 3) outputs. For a 50% duty cycle these pins will produce the standard waveforms they are named for. As the duty cycle gets further away from 50% (in either direction), these waveshapes will become increasingly distorted. At the extremes of the duty cycle range, the triangle wave output will generate a sawtooth wave.

The circuit shown in Fig. 7-9 is certainly not the only possible way to use the 8038, although this arrangement is certainly the most versatile.

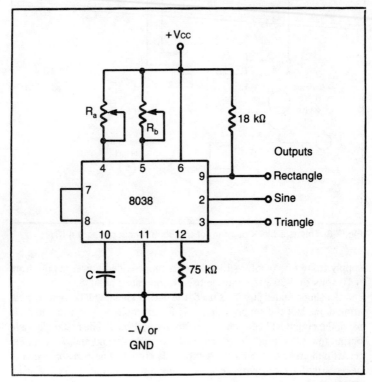

Fig. 7-9. This is the basic arrangement of the timing components for the 8038.

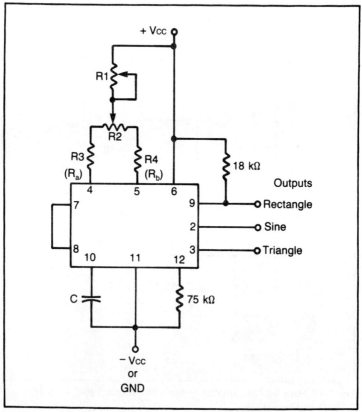

Fig. 7-10. This variation on the basic 8038 offers adjustable frequency and duty cycle.

The operator has full control over both the frequency and the duty cycle via the two potentiometers. On the other hand, this circuit can be a little tricky to use, because both potentiometers must be readjusted to change the frequency while holding the duty cycle constant. Another approach is illustrated in Fig. 7-10. Potentiometer R1 can be used to adjust the frequency. Potentiometer R2 allows the duty cycle to be varied slightly around 50%. Resistors R3 and R4 have equal values.

Probably the simplest possible arrangement is illustrated in Fig. 7-11. The frequency is set by potentiometer R1 which serves as both R_a and R_b. Since the same component is used for both timing resistances, they are, by definition, equal. Therefore, the output will be 50%. The duty cycle cannot be adjusted with this circuit.

Figure 7-12 shows a simple variation that allows you to set up a duty cycle other than 50%. Once again the duty cycle will be constant, and nonadjustable.

117

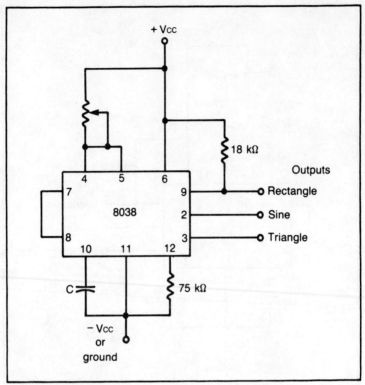

Fig. 7-11. Here we have what is probably the simplest frequency selection scheme for the 8038.

The three timing components (R$_a$, R$_b$, and C) determine the basic nominal output frequency. The actual output frequency can be deviated from its nominal value by feeding a control voltage to the FM (*frequency modulation*) sweep input (pin 8). The 8038 can be used as a VCO (*voltage-controlled oscillator*). In other applications an ac signal will be applied to this input to sweep the function generator through a range of frequencies in a specific, repeating pattern (waveshape). Two 8038's may be cascaded, with one serving as a sweep signal source for the second.

Figure 7-13 shows a practical audio oscillator circuit built around the 8038 function generator. The output frequency can be varied over a 1000:1 range over the complete audible range (20 Hz to 20 kHz). This is done by applying different dc voltages to the FM sweep input (pin 8) via potentiometer R4. Meanwhile, the voltage across the timing resistors (R1 and R3) is held at a relatively low level by the 1N457 diode. The voltage supplied to the timing resistors and the duty-control potentiometer (R3) is several millivolts below the maximum voltage (V$_{CC}$) available to the frequency-control potentiometer (R4). R6 is a trimpot. It is adjusted to minimize variations in the duty cycle with changes in frequency. Potentiometer R7 is adjusted for minimum sine wave distortion at a 50% duty cycle.

The XR-2206 Function Generator IC

Another popular and versatile function generator IC is the XR-2206, which is shown in Fig. 7-14. A block diagram of this chip's internal circuitry is shown in Fig. 7-15. Like the 8038, the XR-2206 generates high quality sine waves, triangle waves, sawtooth waves, and rectangle waves with various duty cycles. The output frequency range runs from a few fractions of a hertz to several hundred kilohertz. A 2000:1 output frequency range is possible with a single variable resistance or control voltage. Power-supply requirements for this chip are very flexible. The XR-2206 can be operated from a single-ended supply from +10 volts to +26 volts, or a dual-polarity power supply from ±5 volts to ±13 volts.

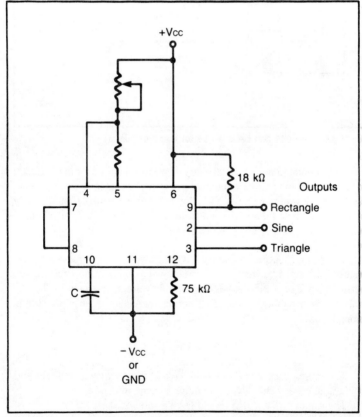

Fig. 7-12. This variation on the circuit shown in Fig. 7-11 allows a duty cycle other than 50%.

119

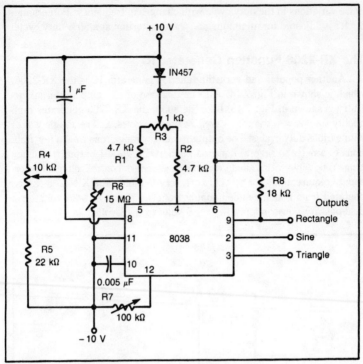

Fig. 7-13. The 8038 can be used as an audio oscillator.

Only a minimum of external components are required for the XR-2206 to operate as a full-featured function generator. Of course, the control voltage input allows frequency modulation (FM) of the output signal, simply by inputting an ac signal. The XR-2206 also has capabilities for AM (*amplitude modulation*), FSK (*frequency-shift keying*), and PSK (*phase-shift keying*).

The nominal output frequency is determined by an external timing capacitor (connected between pins 5 and 6), and a timing resistor (connected between pin 7 or 8 and the negative supply voltage or ground). The formula for determining the nominal output frequency is certainly simple enough:

$$F = 1/R_t C_t$$

The timing capacitor (C_t) should have a value between 1000 pF, and 100 μF. Obviously this gives you quite a range of choice.

The timing resistor (R_t) should have a value in the 1 kΩ (1000 ohms) and 2 MΩ (2,000,000 ohms) range. For the best thermal stability and to minimize distortion of the sine wave, it is best to avoid the extreme values.

For best results keep the value of R_t between 4 kΩ (4000 ohms) and 200 kΩ (200,000 ohms).

The timing resistor may be connected to either pin 7 or pin 8. Why this choice? This allows the use of two separate timing resistors. Switching between the two timing resistors will cause the output to switch between two discrete frequencies. This allows FSK (frequency-shift keying), which is a useful means of communicating encoded data.

Switching between two discrete frequencies can also produce warble-tone effects, which are useful for alarms and other attention-getting applications. If the switching is performed at a very high rate (above 10 to 15 Hz), the two tones will blend together into a single complex composite signal.

The active timing resistor pin (7 or 8) is selected via pin 9. If a voltage greater than two volts is applied to this pin (or if the pin is left open), the pin 7 resistor will be active. The pin 8 resistor is selected by feeding a voltage of less than one volt to pin 9.

The gain and the output phase of the internal multiplier stage can be adjusted by applying a control voltage to pin 1. The output is linearly con-

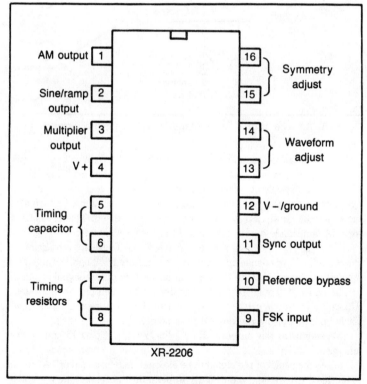

Fig. 7-14. Another popular function generator is the XR-2206.

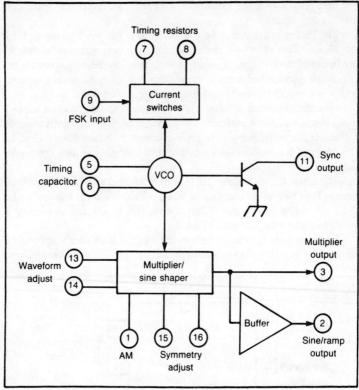

Fig. 7-15. This is a block diagram of the XR-2206's internal circuitry.

trolled by variations around a nominal value equal to one-half the supply voltage. If pin 1 is held at this level, the output will be zero. The gain is increased as the voltage on pin 1 is increased over one-half the supply voltage. Decreasing the pin 1 voltage below one-half the supply voltage also increases the multiplier gain, but the phase is reversed. This pin is used for AM (amplitude modulation) and PSK (phase-shift keying).

There are two main outputs on the XR-2206. These are pins 2 and 3. Pin 3 is a high-impedance output. Pin 2 is a buffered 600-ohm output. The level of the input signal to the internal buffer (the output signal at pin 2) can be varied by connecting a voltage divider between pin 3 and ground. This allows a simple method of gain control. This feature can also be used for keying or pulsing the pin 2 output signal.

The signal at this output will be a linear ramp if pins 13 and 14 are left open. Placing a resistance of a few hundred ohms across pins 13 and 14 causes the peaks of the signal to be exponentially rounded off. At some point this creates a sine wave. With proper adjustment, the sine wave distortion can be made as low as 0.5%.

If the same timing resistance is used to control both charging cycles of the timing capacitor, the internal VCO will generate symmetrical triangle waves and square waves.

An additional rectangle wave output is made available at pin 11. Connecting this output to the FSK input (pin 9) will cause the XR-2206 to automatically switch between the pin 7 and pin 8 timing resistors on alternate half cycles. This allows the circuit to generate a great many unusual waveforms. The IC will generate nonsymmetrical linear ramps (sawtooth waves) and nonsymmetrical pulse (rectangle) waves.

Figure 7-16 shows a sine-wave generator built around the XR-2206 function generator IC. The value of the selected capacitor determines the frequency range the output signal covers:

$$
\begin{aligned}
C &= 1\ \mu F && 10\ \text{Hz - 100 Hz} \\
C &= 0.1\ \mu F && 100\ \text{Hz - 1 kHz} \\
C &= 0.01\ \mu F && 1\ \text{kHz - 10 kHz} \\
C &= 0.001\ \mu F && 10\ \text{kHz - 100 kHz}
\end{aligned}
$$

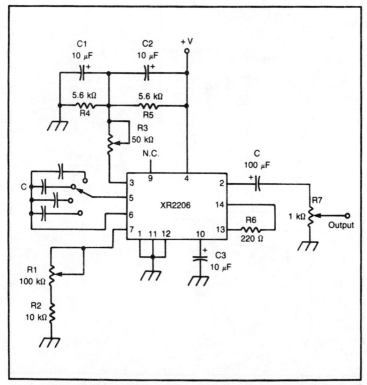

Fig. 7-16. This circuit produces a very clean sine wave with a XR-2206 function generator IC.

123

If just one range is required, of course, just a single capacitor can be permanently wired into the circuit, and the range selector switch can be eliminated.

The timing resistance (R1 and R2) is connected to pin 7. Adjusting potentiometer R1 controls the output frequency. Notice that there is no connection to the FSK input pin (9). When this pin is floating, it acts as if a HIGH signal was applied. Therefore, pin 7 is selected. Potentiometer R3 adjusts the gain of the output signal at pin 2.

Thanks to the 220-ohm resistor between pins 13 and 14, the output is a sine wave. With this simple arrangement, the sine-wave distortion is typically less than 2.5%, which isn't bad.

The circuit shown in Fig. 7-17 generates triangle waves. Notice that it is exactly the same circuit as in Fig. 7-16, except there is no resistor between pins 13 and 14.

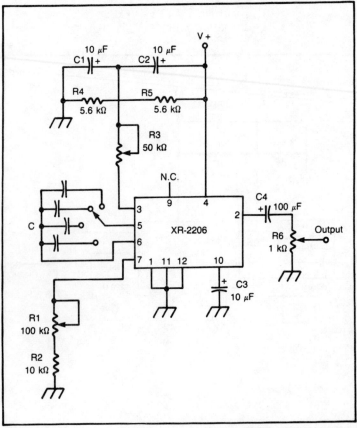

Fig. 7-17. This triangle-wave generator circuit is very similar to the sine wave generator shown in Fig. 7-16.

124

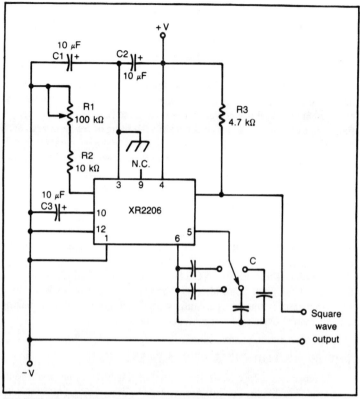

Fig. 7-18. A load/pull-up resistor is required for the square wave output on the XR-2206.

Figure 7-18 shows how a square-wave signal can be generated by the XR-2206. A load/pull-up resistor is connected to the sync output (pin 11), and the output signal is tapped off from this connection. A signal from this output can only be used with high impedance loads (an oscilloscope input, or synchronization terminals, for example). This output cannot be used to drive a low-impedance load. If you must drive a low-impedance load from this circuit, you must use an external buffer stage. The output frequency ranges are the same as for the earlier circuits.

VCO ICs

The signal generator ICs we have discussed so far include VCO (voltage-controlled oscillator) stages. In this section we will look at some chips designed specifically for VCO applications. These devices are commonly used in electronic music and other audio applications.

The design of a VCO in an electronic music system is quite critical.

125

The human ear is very sensitive to changes in frequency. Even someone with no musical training can detect errors as small as 0.06%. Amplitude (volume) is a lot less critical. The smallest change in amplitude the ear can detect is 1 dB, or about 12%. Clearly a VCO must be carefully designed for the maximum possible stability.

The output frequency must follow the control voltage very precisely, according to a specific, consistent scale. Most modern electronic music systems use a standard of one-volt per octave. Note that this is an exponential, rather than linear scale (an octave is a doubling of frequency, so each higher frequency has twice the range as the next lower octave).

An exponential converter is used as an input stage for the control voltage. A transistor's base-emitter voltage to collector current ratio is an exponential relationship, so it is often used to create the converter stage of a VCO. This is all well and good, except for one problem—transistors are temperature sensitive. The control voltage will be affected by changes in temperature, so the output frequency will drift, which we definitely don't want. Fortunately, circuit designers have developed a number of tricks for minimizing thermal instability to acceptable levels.

Two high-tech dedicated VCO ICs are the Curtis CEM3340, and the SSM2030 from Solid State Micro Technology. Both of these devices are very stable and distortion free. Many special features are offered, making a great many specialized applications possible.

THE 50240 TOP OCTAVE SYNTHESIZER IC

Another interesting IC designed primarily for electronic music applications is the 50240 top octave generator. The pinout diagram for this device is shown in Fig. 7-19. This IC accepts a high-frequency clock input signal, and divides it down to 13 audio frequencies, representing the notes in a standard equally tempered octave. The term *equally tempered* refers to a specific type of tuning. It is the type of tuning usually used for such instruments as the piano. Each adjacent note represents a frequency change factor of the twelfth root of two.

The nominal clock frequency for the 50240 is 2.0024 MHz (2,002,400 Hz). This gives an output scale that will be in tune with standard concert instruments. The divisors used to derive each note are as follows:

C_8	pin	16	c/478
$C^\#_8$	pin	4	c/451
D_8	pin	5	c/426
$D^\#_8$	pin	6	c/402
E_8	pin	7	c/379
F_8	pin	8	c/358
$F^\#_8$	pin	9	c/338
G_8	pin	10	c/319
$G^\#_8$	pin	11	c/301

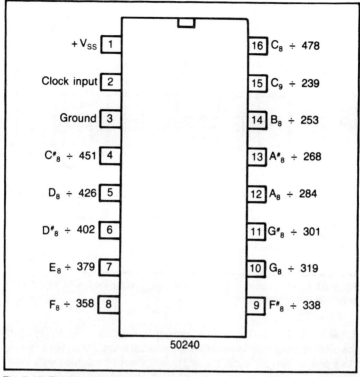

Fig. 7-19. The 50240 top-octave generator IC produces 13 equally tempered notes.

A_8	pin 12	c/284
$A\#_8$	pin 13	c/268
B_8	pin 14	c/253
C_9	pin 15	c/239

where c is the clock frequency.

The number following the note name (8 or 9), refers to the octave number. Octave 8 is generally the highest octave used in music. Since an octave is a doubling of frequency, lower octaves for each note can be achieved simply by dividing each note output by two. This can be done with simple external flip-flop stages, as illustrated in Fig. 7-20.

The 50240 generates the highest useful octave. This is why it is called a top octave generator. The top octave approach has been used in electronic organs for decades. Formerly, top octave generators required complex and expensive circuitry. The 50240 has certainly changed all that. It is inexpensive and quite easy to work with.

The clock frequency can be anything from 100 kHz (100,000 Hz) to

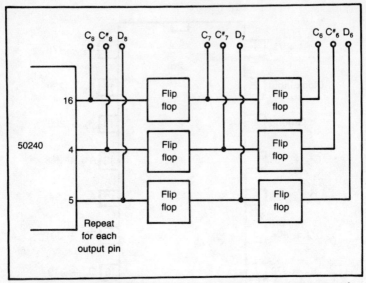

Fig. 7-20. Lower octaves can be obtained by dividing the 50240's output frequencies with external flip-flops.

2.5 MHz (2,500,000 Hz). A clock frequency of 2.0024 MHz will result in the outputs being in tune with standard musical instruments. Even if other clock frequencies are used, the 50240 will always be in tune with itself. Adjacent notes are always separated from each other by an equally tempered twelfth root of two.

The 50240 is very easy to work with. A single-ended power supply is used. The supply voltage can be anything from +11.0 volts to +16.0 volts. The output signals are square waves with 50% duty cycles. This makes them ideally suited for digital division with standard flip-flops. Moreover, a square wave is rich in harmonics, so a variety of different voicings can be created by simple filtering.

Figure 7-21 shows a simple clock generator that can be used to drive the 50240 top octave generator. Potentiometer R2 can be adjusted so the clock is putting out exactly 2.0024 MHz. If a VCO is used as the clock (as shown in Fig. 7-22), the output frequencies of the 50240 will essentially be voltage controlled, but they will always maintain their equally tempered relationship.

THE MM5837 DIGITAL NOISE GENERATOR

It might at first seem that a circuit whose sole function is to produce noise would represent an ultimate level of uselessness. After all, in most circuits, great efforts are taken to keep noise to a minimum. A controllable noise source can be extremely useful in sound synthesis. Noise is used

128

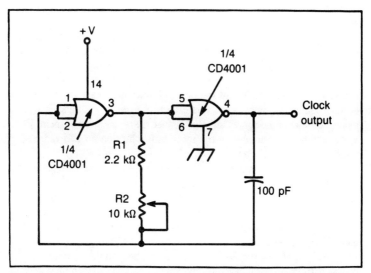

Fig. 7-21. This simple clock circuit is ideal for driving the 50240 top-octave generator IC.

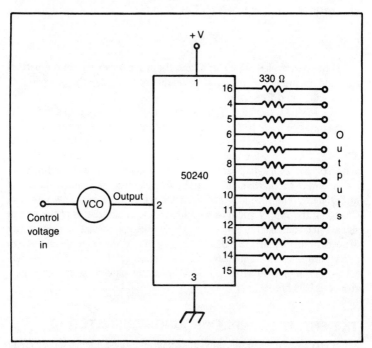

Fig. 7-22. Some interesting results can be obtained by using a VCO as the clock for a 50240 top-octave generator.

to produce such sounds as drums, wind, explosions, and the sound of a wind player's breath, among others.

In this sense, noise is a random (unpredictable, and nonrepeating) varying signal. The most basic type is white noise. At any given instant any possible frequency has an equal chance of being present. No frequency, or group of frequencies is more likely to occur than any other. Since each octave involves a doubling of frequency, higher octaves contain more discrete frequencies. Because of this effect, the upper frequencies will appear to dominate the sound, simply because there are more of them. White noise sounds very much like the interstation hiss on an FM radio when you tune between channels.

Another common type of noise is pink noise. Pink noise has equal energy per octave, rather than per frequency. The odds of the instantaneous frequency being within a given octave are equal for all octaves. Essentially, the lower frequencies are emphasized, and given a greater chance of occurrence. Pink noise can be created by passing white noise through a low-pass filter to deemphasize the higher frequency components.

Ideally, a noise generator's output should be completely random. That is, the pattern of instantaneous frequencies should never repeat itself. It is generally more practical to set up a pseudorandom generator, especially if digital circuitry is used. A pseudorandom generator produces a long jumbled pattern, which is repeated. For example:

$$592623859262385926385 \ldots$$

If the pattern is long enough, and the frequencies within the pattern are random enough, the ear will be unable to recognize the pattern. The result will sound like noise.

The MM5837 (sometimes numbered S2668) is a single chip digital noise generator. The noise pattern produced by this IC is pseudorandom. This IC is one of the easiest there is to use. Only four of the eight pins are actually used (there is a ground connection, two supply voltage connections, and an output terminal). That's it! The pinout diagram is shown in Fig. 7-23.

A simple circuit for simulating the sound of a snare drum is illustrated in Fig. 7-24. The MM5837 can accept a wide range of supply voltages. Changing the supply voltages will alter the characteristics of the output signal somewhat, but the effect generally isn't too significant for the majority of applications. Figure 7-25 illustrates how to feed the noise signal through a low-pass filter to create a pink-noise generator. The MM5837 is ideal for use in conjunction with the MM5871 rhythm pattern generator IC (discussed in Chapter 9).

THE SN76477 COMPLEX SOUND GENERATOR IC

If you need to generate complex signals and sound effects, the SN76477 will be a valuable chip for you. This is a complete sound synthesizer sys-

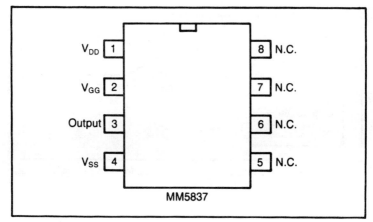

Fig. 7-23. The MM5837 generates a pseudorandom noise signal.

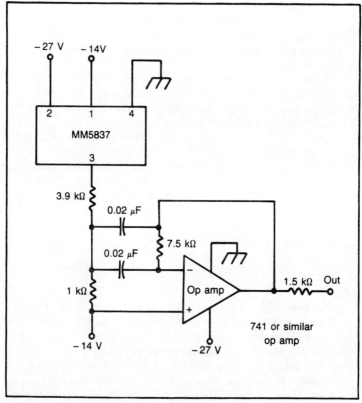

Fig. 7-24. A noise generator like the MM5837 is useful for generating percussive sounds, like that of a snare drum.

131

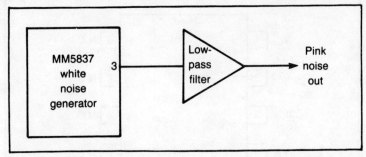

Fig. 7-25. Passing the output of the MM5837 white noise generator through a low-pass filter simulates pink noise.

tem in a 28-pin IC package. Only a handful of external resistors and capacitors (and possibly some switches) are required to generate any of thousands of possible sounds.

While the SN76477 could conceivably be used for electronic music synthesis, it is primarily intended to generate sound effects (it is used in a number of video games, for example), and really isn't very well suited for most musical applications.

The pinout diagram for this device is shown in Fig. 7-26. Figure 7-27 is a block diagram of the SN76477's internal structure. Because most of the pin functions on this chip are likely to be unfamiliar to many readers, we will examine the SN76477 pin-by-pin.

Pin 1—This pin, along with pin 28, determines the envelope select logic. External logic signals (HIGH or LOW) are applied to these two pins to select the envelope function. There are four possible combinations:

Pin 1	Pin 28	Function
0	0	VCO
0	1	Mixer only
1	0	One-shot
1	1	VCO with alternating polarity

Pin 2—This pin is simply the ground connection for the chip.

Pins 3 and 4—The noise clock controls the noise generator. A 43 kΩ to 50 kΩ resistor is connected from pin 4 to ground for this circuit to function. Alternatively, an external noise clock signal may be applied to pin 3. The input signal should not have an amplitude greater than 5 volts peak-to-peak. It should be a square wave for proper operation.

Pins 5 and 6—The SN76477 includes a binary pseudorandom white-noise generator. This noise signal is passed through a variable bandwidth low-pass filter. Filtering at various frequencies will cause the noise to sound

132

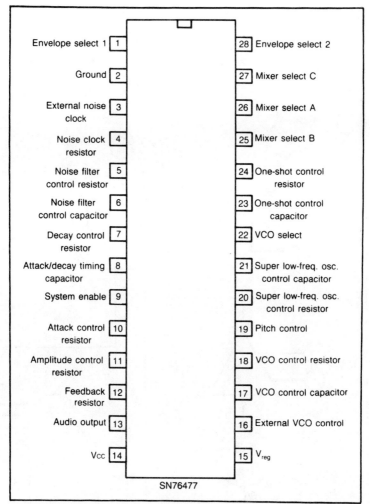

Envelope select 1	1	28	Envelope select 2
Ground	2	27	Mixer select C
External noise clock	3	26	Mixer select A
Noise clock resistor	4	25	Mixer select B
Noise filter control resistor	5	24	One-shot control resistor
Noise filter control capacitor	6	23	One-shot control capacitor
Decay control resistor	7	22	VCO select
Attack/decay timing capacitor	8	21	Super low-freq. osc. control capacitor
System enable	9	20	Super low-freq. osc. control resistor
Attack control resistor	10	19	Pitch control
Amplitude control resistor	11	18	VCO control resistor
Feedback resistor	12	17	VCO control capacitor
Audio output	13	16	External VCO control
Vcc	14	15	V_{reg}

SN76477

Fig. 7-26. The SN76477 complex sound generator IC can produce hundreds of different sounds.

quite different. The 3 dB cut-off frequency (F_c) is determined by the value of an external resistor connected from pin 5 to ground (R_{nf}) and an external capacitor from pin 6 to ground (C_{nf}). The approximate cut-off frequency can be estimated with this formula:

$$F_c = 1.28/R_{nf}C_{nf}$$

Pins 7, 8, and 10—These pins are used to set the timing values for the envelope generator. Most sounds don't just switch on, instantly jump-

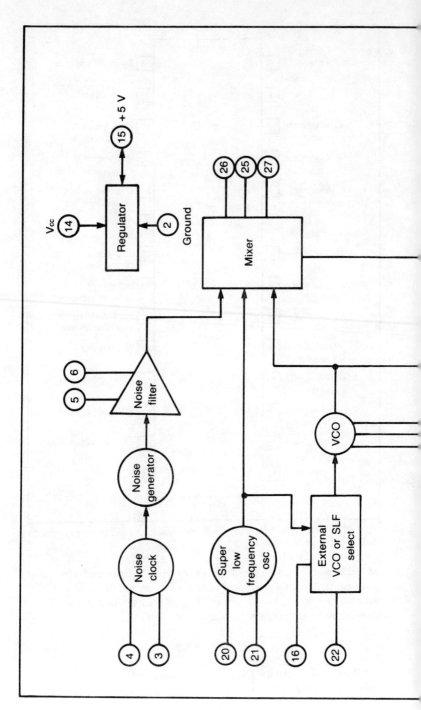

134

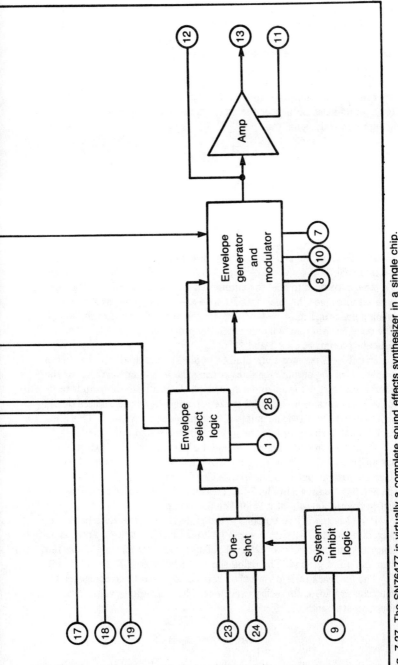

Fig. 7-27. The SN76477 is virtually a complete sound effects synthesizer in a single chip.

ing to their maximum amplitude (instantaneous volume level), and then switch instantly back off to zero. Rather, they take some finite time to build up to the maximum level (attack), and then again to drop back down to zero (decay). The envelope generator allows control of these amplitude changes with time.

A common timing capacitor (C_{ad}) is used for both the attack and decay times. It is connected from pin 8 to ground. A resistor (R_a) from pin 10 to ground also influences the attack time. The attack time in seconds is approximately equal to:

$$T_a = R_a C_{ad}$$

Similarly, the decay time is determined by the capacitor at pin 8, and a resistor (R_d) which is placed between pin 7 and ground:

$$T_d = R_d C_{ad}$$

Pin 9—This pin is used to inhibit or enable the chip system. A logic 0 (LOW) enables the system. A logic 1 (HIGH) inhibits it. The inhibit/enable pin can be used to hold the output in a no sound condition. It can also be used to trigger the one-shot. The one-shot is triggered by a negative-going pulse. Pin 1 must be held LOW for the entire duration of the sound. The one-shot will function only when the proper envelope select logic inputs are present at pins 1 and 28.

Pin 11—A resistor connected from pin 11 to ground sets the overall amplitude of the output signal, by setting the operating currents for the internal circuitry of the op-amp output stage. Typically, the amplitude resistor will have a value in the 47 kΩ to 220 kΩ range. Lower resistances should be avoided, because the op amp will tend to become saturated. This is especially noticeable on the decay portion of the sound envelope.

Pins 12 and 13—These two pins are used as part of an external output amplifier. The SN76477 is not designed to drive a speaker directly, so an external amplifier is required. A simple typical amplifier for use with the SN76477 is shown in Fig. 7-28. This amplifier will drive a small 8-ohm speaker with 300 to 400 mW (0.3 to 0.4 watt).

Pin 13 is the actual output for the chip. A feedback resistor is connected to pin 12. The output at pin 13 is an emitter-follower. There is no internal load resistor, so an external pull-down resistor must be connected from pin 13 to ground. The value should be between 2.7 kΩ and 10 kΩ.

The feedback resistor, which is connected to pin 12 compensates for external, and any chip-to-chip variations. The peak output voltage can be approximated with this formula:

$$V_o = 3.4 R_f / R_{amp}$$

where R_f is the feedback resistor connected to pin 12, and R_{amp} is the am-

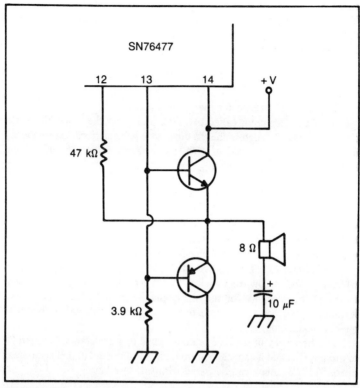

Fig. 7-28. This simple amplifier can be added to the SN76477 to drive a small speaker.

plitude resistor at pin 11.

The dynamic output range of the SN76477 is limited to 2.5 volts peak-to-peak. If you try for a larger output voltage, clipping distortion will appear.

Pin 14—The supply voltage is normally applied to this pin (see also pin 15). The supply voltage at pin 14 can be anything from +7.5 volts to +12 volts, although +9 volts is the recommended maximum.

Pin 15—This pin is rather unusual. It can be used as a regulated voltage output or input. The SN76477 contains an internal +5.0-volt regulator that is available at pin 15. It can be used as a control voltage source for the chip's other inputs. The internal regulator can supply up to 10 mA (0.01 amp) of current.

Alternatively, an external +5.0 volt regulated power supply can be used to drive the chip. The external regulated voltage is fed to pin 15, and the ordinary supply voltage pin (14) is left unused.

Pin 16—An external control voltage may be optionally applied to pin 16 to control the pitch (frequency) of the VCO.

Pins 17 and 18—The nominal frequency of the VCO is set by a re-

137

sistance from pin 18 to ground, and a capacitance from pin 17 to ground. The nominal frequency is the minimum frequency, which is produced when the control voltage (whether from an internal or external source) is zero. The formula for approximating the nominal output frequency of the VCO is:

$$F = 0.64/R_{vco}C_{vco}$$

The frequency range for the VCO is about 10:1.

Pin 19—This pin is called the pitch control for the VCO. More specifically, it is used to determine the duty cycle (which affects the frequency indirectly). The output of the VCO is a rectangle wave. The duty cycle is determined by this formula:

$$DC = 50 \times V_{16}/V_{19}$$

where V_{16} is the voltage at pin 16, and V_{19} is the voltage at pin 19. (This formula gives the duty cycle in percent.) A constant 50% duty cycle (square wave) can be achieved by holding pin 19 HIGH.

Pins 20 and 21—The SN76477 also contains a second oscillator, which is set up for low frequencies. This is the *super low-frequency oscillator* (SLF). It is designed for output frequencies from 0.1 Hz to 30 Hz, but it can be forced into operation up to 20 kHz, allowing all sorts of FM (frequency modulation) effects.

The frequency of the SLF is determined by a resistance from pin 20 to ground (R_{lf}) and a capacitance from pin 21 to ground (C_{lf}). The approximate SLF frequency can be found with this formula:

$$F_{lf} = 0.64/R_{lf}C_{lf}$$

The super low frequency oscillator produces two waveforms. A square wave is fed to the mixer (discussed shortly). In addition, a triangle wave is fed to the VCO control input when pin 22 is HIGH.

Pin 22—This pin is used to select the control voltage source for the VCO. If pin 22 is held LOW, an external control voltage (applied to pin 16) drives the VCO. On the other hand, holding pin 22 HIGH allows the super low frequency oscillator (SLF) to control the VCO.

Pins 23 and 24—These two pins are used to set the timing period of an internal one-shot stage, or monostable multivibrator (see also Chapter 3). The timing capacitor (C_{os}) is placed between pin 23 and ground. The timing resistor (R_{os}) is connected between pin 24 and ground. The approximate timing period for the one shot (in seconds) is:

$$T = 0.8R_{os}C_{os}$$

The one-shot is used for non-continuous sounds. A typical example would be a gunshot.

Pins 25, 26, and 27—These three pins are logic inputs that control the external mixer stage. The states of these control pins determine which of the several internal signals in the SN76477 will be applied to the output.

Pin 26 is mixer control A, pin 25 is mixer control B, and pin 27 is mixer control C. There are eight possible combinations of digital inputs for these three pins. Each combination results in a different output (or combination of outputs):

A	B	C	output
0	0	0	VCO
0	0	1	SLF
0	1	0	Noise
0	1	1	VCO/Noise
1	0	0	SLF/Noise
1	0	1	VCO/SLF/Noise
1	1	0	VCO/SLF
1	1	1	Inhibit (no output)

As you can see, the SN76477 is designed to create a wide variety of different sounds.

Pin 28—See the information on Pin 1.

For the best results, each of the various programming resistors and capacitors should be kept within certain ranges. Each of the component values ideally should be within the limits outlined below:

Resistors

Pin 4	40 kΩ - 50 kΩ
Pin 5	7.5 kΩ - 1 MΩ
Pin 7	7.5 kΩ - 1 MΩ
Pin 10	7.5 kΩ - 1 MΩ
Pin 18	7.5 kΩ - 1 MΩ
Pin 20	7.5 kΩ - 1 MΩ

Capacitors

Pin 6	150 pF - 0.01 μF
Pin 8	0.01 μF - 10 μF
Pin 17	100 pF - 1 μF
Pin 21	500 pF - 100 μF
Pin 23	0.1 μF - 50 μF

The SN76477 complex sound generator IC is a great device for the experimenter, because there are so many things you can do with it. Feel free to try out any idea that might occur to you. This device is pretty hard to damage as long as you don't exceed the input voltage limitations. Generally speaking, as long as you don't apply more than 6 volts to any of the input pins (except power supply pin 14, of course) you can be reasonably sure you won't do any harm. Usually the worst that will happen is that you'll get a terribly obnoxious sound, or no sound at all. In which case, try something else. Believe me, you'll never go through all the possible circuits for the SN76477.

The SN76477 certainly can't do everything. It generally falls short of the requirements for serious electronic music applications. But it can sure do a lot! We'll look at just a few possibilities, before moving on to other ICs.

The noise source is a pseudorandom digital circuit. The clock rate will influence the nature of the sound in some fairly subtle, but definite ways. Ordinarily the clock frequency is fixed by a simple resistor between pin 4 and ground. But you can get fancier, if you like. For example, try connecting a transistor to pin 4, as shown in Fig. 7-29. This arrangement allows dynamic control of the noise clock frequency, via the signal applied to the base of the transistor. Some fascinating "swooshing" effects can be achieved if the base is connected via a large value resistor to pin 8, which is the connection point for the envelope timing capacitor. The voltage at this point varies as the capacitor charges and discharges. The sound quality of the noise will vary along with the envelope.

This trick of substituting a transistor for a resistor can be used for any of the programming resistors. In some cases the results won't be particularly useful, but in others, you can create some very novel sounds. Remember, a transistor is basically a variable resistance element. The very name comes from *trans*fer res*istor*.

The SN76477's VCO is intended to only produce square waves, but you can trick it into generating triangle waves. A partial circuit for accomplishing this is shown in Fig. 7-30. There is one major limitation with this

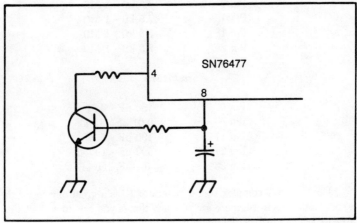

Fig. 7-29. Connecting a transistor to pin 4 of the SN76477 allows dynamic control of the noise clock frequency.

140

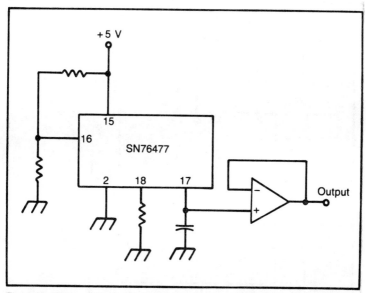

Fig. 7-30. This partial circuit can be used to get a triangle wave output from the SN76477.

approach. The signal amplitude is sensitive to changes in the control voltage applied to pin 16. In other words, the volume changes with frequency. This doesn't seem to occur with the regular square wave output, or when the internal voltage control is used.

Two typical circuits using the SN76477 are presented in Figs. 7-31 and 7-32 to get you started on your experimenting. Using the component values listed, the circuit in Fig. 7-31 will produce a sound reminiscent of a laser gun in a science fiction movie. The component values listed in Fig. 7-32 will give a gunshot effect. Of course you can create a wide variety of other sounds with these circuits, simply by changing one or more of the component values.

Closely related to the SN76477 is the SN76488. This is an improved version with an on-chip low-power output amplifier that can drive a small speaker directly.

THE SN94281 COMPLEX SOUND GENERATOR

Another complex sound generator IC is the SN94281. It is a more recently developed device than the SN76477. It is much easier to use than its predecessor. Despite the fact that it has only 16 pins (as compared to the 28 pins of the SN76477) it is every bit as versatile, if not even more so. The pinout diagram for the SN94281 is shown in Fig. 7-33. This chip's internal circuitry includes both analog and digital technology. It can be used with just a few external resistors, capacitors, and switches, or it can be

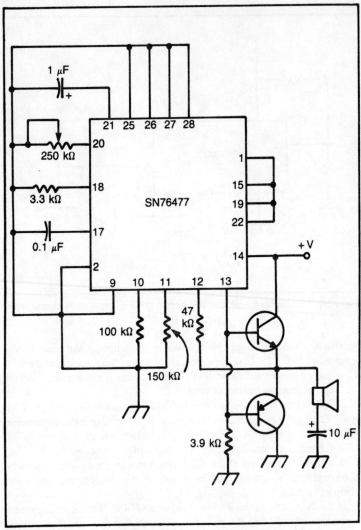

Fig. 7-31. This circuit can be used to create a laser gun sound.

interfaced directly to a microcomputer.

A built-in 125 mW (0.125 watt) amplifier is included on the chip. It can drive a small speaker directly, or it can be used as an input to a more powerful amplifier. Figure 7-34 shows a typical circuit built around the SN94821. Using the component values shown in the schematic, the effect will be like a steam locomotive. Momentarily closing the switch will cause a whistle to be sounded. Of course, different component values (and configurations) can produce a wide variety of additional sounds.

THE AY-3-8910/8912
PROGRAMMABLE SOUND GENERATOR ICs

When the SN76477 first came out, everyone was justifiably impressed by how easy it was to create thousands of different sounds with just a few external resistors and capacitors. The AY-3-8910/8912 Programmable Sound Generator ICs go one step better—all sounds are digitally created, eliminating the need for external components altogether.

These devices are designed to be operated under computer control, and they are incredibly versatile. Three independent channels are provided, allowing the creation of very complex sounds. Each of the channels has

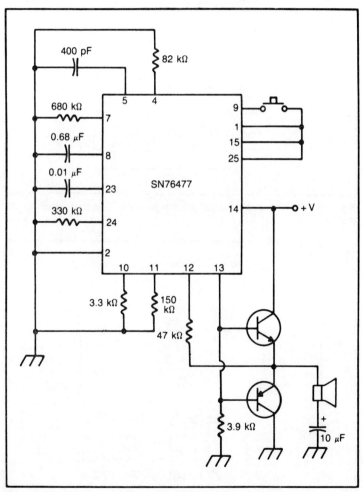

Fig. 7-32. Using the component values shown, this circuit will produce a gunshot effect.

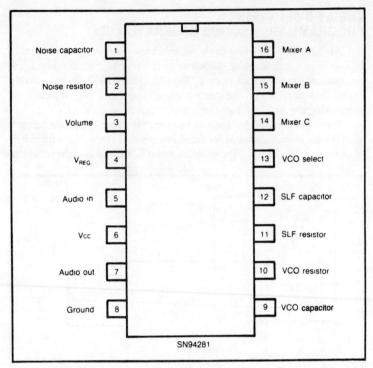

Fig. 7-33. The SN942081 complex sound-generator IC is even easier to use than the SN76477.

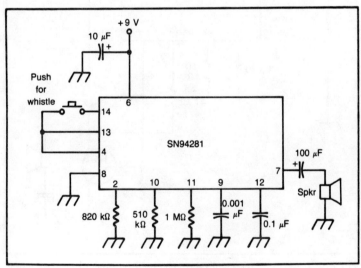

Fig. 7-34. This circuit simulates the sound of a train.

144

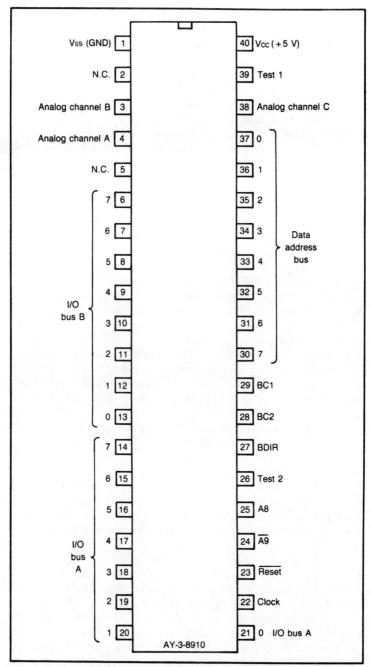

Fig. 7-35. The AY-3-8910 is a powerful programmable sound generator.

145

an analog output from a 4-bit logarithmic D/A (digital-to-analog) converter. There are two versions of this device. The AY-3-8910 is contained in a 40-pin DIP housing, as shown in Fig. 7-35. It features two 8-bit parallel ports.

The AY-3-8912 is a slightly stripped down version. It is housed in a 28-pin DIP package. The pinout diagram is given in Fig. 7-36. The primary difference between the AY-3-8912 and the AY-3-8910 is that the AY-3-8912 has only a single 8-bit parallel port. A block diagram of the internal structure of these chips is shown in Fig. 7-37.

For convenience, we will confine the rest of our discussion to the AY-3-8910. For the most part, the same basic information will also apply to the smaller AY-3-8912.

The *programmable sound generator* (PSG) requires only a single command for each sound parameter. Internal latches hold the commands, so that a sound may be held as long as desired, with no further attention from the microprocessor. The computer only has to address the PSG when some parameter of the sound is to be changed. This saves a considerable amount of memory space and program time. The computer can simultaneously perform a number of other tasks as the sound is being produced by the PSG.

The AY-3-8910 is ideal for almost any application where a computer-controlled (or other digital signal source) audio signal is required. Electronic music and sound effects generation are the most obvious types of use, but alarm systems, FSK (frequency-shift keying) communications, and tone signaling are additional possibilities.

This chip has features galore. Three wide-range square-wave generators are provided. Depending on the clock frequency used, the frequency of these generators can range from sub-audio to ultra-audio tones. A fourth signal source in the AY-3-8910 is a pseudorandom digital noise source for percussive type effects.

The AY-3-8910 has three independent signal channels. Each channel has its own mixer for combining the outputs of the tone generators and noise source in various combinations. Either a fixed or variable amplitude pattern can be applied to each individual channel. The fixed amplitude pattern is taken from the digital data from the controlling computer. The variable pattern can be set up with an internal envelope generator. Both the shape and the cycle of the envelope are user programmable. The I/O ports are used for interfacing the host computer with other devices, in addition to the AY-3-8910.

The AY-3-8910 has sixteen memory-mapped internal registers to control the various parameters of the sound generation/modification stages, and the I/O ports. To the controlling computer, the registers appear as a 16-word block out of 1024 possible addresses.

The envelope generators can produce 10 distinct patterns, under the control of four bits. These patterns are illustrated in Fig. 7-38. The frequency generated by the tone generators is dependent upon the clock frequency driving the chip (F_c), and the controlling digital data in the

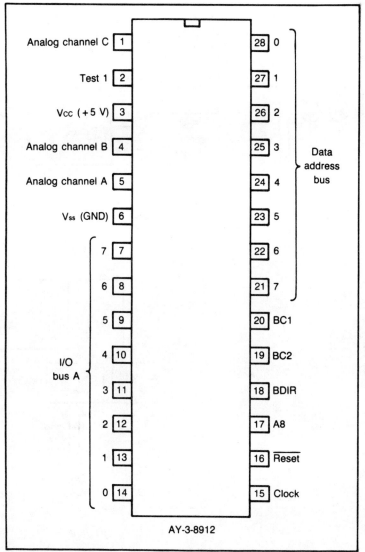

Fig. 7-36. The AY-3-8912 is a slightly simplified version of the AY-3-8910.

appropriate register (D). The formula for the tone generator output frequency works out to:

$$F = F_c/16D$$

The parameter defining commands are stored in several internal registers (small memory banks). Each register has a specific function.

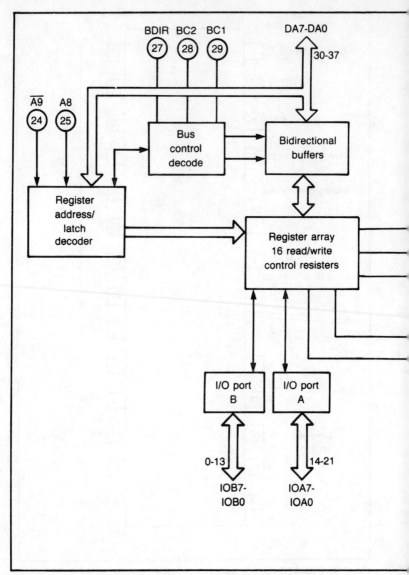

Fig. 7-37. This block diagram summarizes the internal circuitry of the AY-3-8910 programmable sound-generator IC.

Two of the registers (R0 and R1), for example, combine to determine the tone period for the tone generator in channel A. The tone period of a signal is the reciprocal of its frequency:

$$TP = 1/F$$

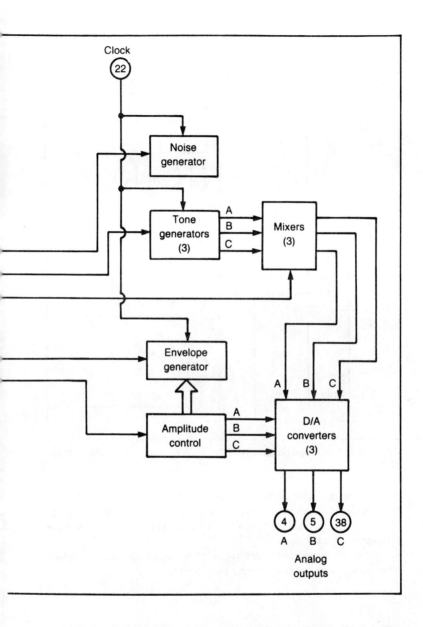

R0 uses 8 bits for fine tuning (256 steps). R1, on the other hand, is a 4-bit (16 steps) coarse tuning control. Basically, R1 sets the approximate range, and R0 sets the exact tone period. The twelve bits of R0 and R1 together makes for an impressively wide range of tone periods.

The tone period is created by first dividing the system clock frequency

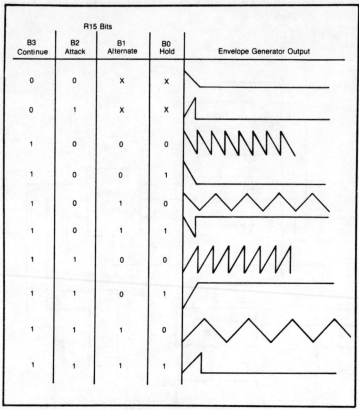

R15 Bits				
B3 Continue	B2 Attack	B1 Alternate	B0 Hold	Envelope Generator Output
0	0	X	X	
0	1	X	X	
1	0	0	0	
1	0	0	1	
1	0	1	0	
1	0	1	1	
1	1	0	0	
1	1	0	1	
1	1	1	0	
1	1	1	1	

Fig. 7-38. The envelope generators of the AY-3-8910 can produce ten distinct patterns.

by 16. Then this value is divided by the 12-bit number defined by R1 and R0. This number can range from a low of 000000000001 (highest possible frequency—divided by 1) to a high of 111111111111 (lowest possible frequency—divided by 4,095).

If we call the decimal equivalent of the register contents P, the output frequency can be calculated using this formula:

$$F = C/16P$$

where C is the original system clock frequency, and F is the output frequency.

Similar register pairs define the tone periods for the tone generators in the other two channels.

The output signals from all the AY-3-8910's tone generators are always square waves. This admittedly puts some limitations on the possible sounds that can be created by the PSG. But it is not as big a limitation as it may seem at first. Other waveforms can be simulated by combining channels, external filtering, or rapidly varying the tone period of one or more of the tone generators for a frequency modulated (FM) effect.

Another register in this chip (R7) is used to enable or disable various portions of the PSG. If bit B0, for instance, is a logic 0, the tone is enabled in channel A. A logic 1 for B0 disables this tone. Similarly, bit B1 controls the tone in output channel B, and bit B2 controls the tone in output channel C.

Bits B3, B4, and B5 enable/disable the noise source in each of the three channels. Again a logic 0 enables the noise source, and a logic 1 disables it.

Bit B6 controls the input/output (I/O) port. A logic 0 in this position means the port is in the input mode. A logic 1 in B6 puts the port in the output mode. Bit B7 of register R7 is not used in the 28-pin AY-3-8912. In the 40-pin AY-3-8910, this bit controls a second I/O port in the same manner as B6.

The two I/O ports can be interfaced with an external memory, as illustrated in Fig. 7-39. One port is left in the output mode for memory addressing, and the other port is used for data input/output.

It should be mentioned here that using register R7 to disable both the tone and the noise source through a given channel does not turn that channel off. That must be done by loading all 0's in the appropriate amplitude (volume level) register. All three output channels of the PSG may be fed to a single amplifier and speaker. Alternatively, each channel may be fed to its own individual amplifier and speaker.

A great many special effects are possible with the Programmable Sound Generator. As with most sound synthesis devices, the number of possible sounds is really limited primarily by the user's imagination and ingenuity.

Simple one-note melodies are obviously quite easy to produce. For greater musical interest, two or three of the output channels can be simul-

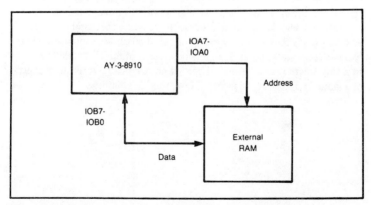

Fig. 7-39. The I/O ports of the AY-3-8910 can be interfaced with external memory.

151

taneously used for harmonic effects. Since we have three independent voices available, we can conveniently produce chords. The three channels can also be blended together to generate different tonal qualities. By combining two or three square waves of different frequencies and with differing amplitude patterns, many other usual waveshapes may be created.

This powerful device can be the heart of an extremely versatile sound synthesis system. There are any number of ways of telling the microprocessor/PSG system what you want it to do. One popular approach is to use an organ-like keyboard that can be digitally scanned to determine which (if any) keys are being held down at any instant. Sensors for key-strike velocity and depth can also be used. This digitally scanned keyboard system is used in most modern polyphonic synthesizers.

Data may be entered so that the sound is instantly produced (as with most ordinary musical instruments—you hear what you're playing as you play it). Alternatively, data can be stored for later playback. The most powerful systems, of course, would offer both options.

If the data is stored in the microprocessor's memory for later playback, the tune (or sequence of sounds) can be manipulated in a variety of ways. For instance, a relatively simple program could raise or lower the entire piece an octave, or transpose it to another key.

The tempo, or speed, of a passage may also be manipulated. If a given sound event was originally programmed to last 20 clock pulses, it could easily be changed to last 10 clock pulses (speeded up), or 40 clock pulses (slowed down), or almost any value desired. This would not affect the pitches of the tones or any other parameter of the sound, other than the tempo. That's the big advantage of the PSG right there. Each parameter of the sound is independently controllable. They are not affected by each other in any way.

The AY-3-8910 is quite a complex device, and it would not be appropriate to go into greater detail here. A full examination of this chip and its capabilities would fill a book of its own. The manufacturer, General Instruments, offers an excellent 32 page data sheet/manual.

If the SN76477 and the SN94281 can generate thousands of different sounds (which they certainly can), the AY-3-8910 can generate millions. Since it is computer controlled, it is possible to jump from sound to sound in a tiny fraction of a second. There is no need to re-wire the circuitry, or manually adjust any controls. Now, that's versatility.

Chapter 8

Sound Modifier ICs

I N THE LAST CHAPTER, WE LOOKED AT SIGNAL GENERATOR ICS, WHICH
are used to create audible sounds. In this chapter we will examine some
ICs that are designed to modify signals in various ways, primarily for audio applications.

FILTERS

Most ac (audio) signals contain many frequency components. Only the
sine wave is a pure, single frequency signal. Periodic (repeating) signals
(such as sawtooth waves, rectangle waves, etc.) include a number of frequency components called harmonics, which are integer multiples of the
main (fundamental) frequency of the waveshape.

A filter is a circuit that allows certain frequency components to pass
through to its output, but blocks other frequency components. There are
four basic types of filters, classified according to their pattern of passed
and blocked frequencies.

☐ A *low-pass filter* passes low frequencies, as the name suggests. High
frequencies are blocked. The dividing point (or cut-off frequency) is usually
labelled F_c. The frequency response of an ideal low-pass filter is graphed
in Fig. 8-1.

☐ A *high-pass filter* is just the opposite of a low-pass filter. Low frequencies are blocked and high frequencies are passed. The frequency response graph, shown in Fig. 8-2, is the inversion of the low-pass graph.

☐ A *band-pass filter* is slightly more complex. In this type of filter,
a specific band of frequencies is passed. Any frequency components above
or below that band are rejected (blocked). The frequency response graph

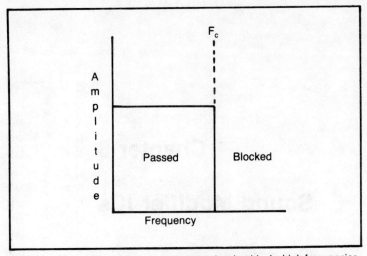

Fig. 8-1. A low-pass filter passes low frequencies, but blocks high frequencies.

for an ideal band-pass filter is shown in Fig. 8-3. Two specifications are needed to define the action of a band-pass filter. The center frequency (also labelled F_c) is the mid-point of the passed band. The bandwidth (BW) is the range of frequencies contained within the passed band. For example, if a band-pass filter passes only those frequencies between 2000 Hz and 3000 Hz, the center frequency is 2500 Hz, and the bandwidth is 1000 Hz.

☐ A *band-reject filter* is the opposite of a band-pass filter. All frequencies are passed except for those within the specified band. Again, the action of the filter is defined by the center frequency (mid-point of the rejected band), and the bandwidth. A frequency response graph for an ideal band-

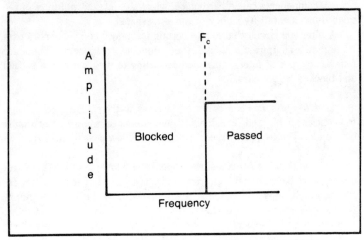

Fig. 8-2. A high-pass filter is just the opposite of a low-pass filter.

154

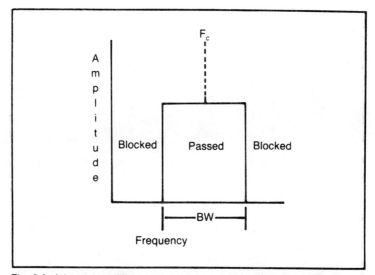

Fig. 8-3. A band-pass filter passes only a specific range of frequencies.

reject filter is shown in Fig. 8-4. Because of the appearance of its frequency response graph, the band-reject filter is sometimes called a notch filter. The terms are interchangable.

The frequency response graphs shown in Figs. 8-1 through 8-4 are idealized. All frequency components are either completely passed or completely blocked. This just isn't possible in the real world. Practical filter circuits have a middle-ground around the cut-off frequency (or edges of the pass band) in which frequency components are increasingly attenuated. Figure 8-5 shows the frequency response graph of a practical low-pass filter. The

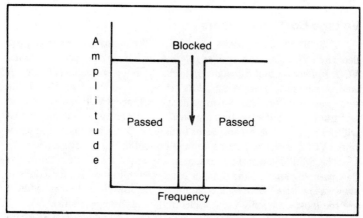

Fig. 8-4. A band-reject filter is also sometimes known as a notch filter.

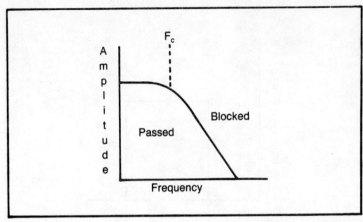

Fig. 8-5. A practical filter has a finite slope between the passed and blocked frequencies.

cut-off slope is measured in x dB per octave. Obviously, the steeper the slope (the higher the value of x), the more precise the operation of the filter.

Passive Filters

Simple filters can be constructed from passive components. For example, Fig. 8-6 shows a simple low-pass filter made up of just a resistor and a capacitor. Passive filters are not sufficient for many applications. They attenuate all frequencies somewhat and the cut-off slope is very gradual.

Better filters can be constructed with active stages. Several dedicated filter ICs have been developed. These ICs have a number of interesting features, which give them a great deal of versatility.

Voltage-Controlled Filters

The SSM2040 from Solid State Micro Technology, for example, is a dedicated VCF (*voltage-controlled filter*). A VCF is similar in concept to the VCOs (*voltage-controlled oscillators*) discussed in the preceeding chapter. A control voltage adjusts the cut-off (or center) frequency of the filter. If the same control voltage is used to drive both a VCO and a VCF, the two devices will track each other, so that the harmonic content of the output signal will remain constant, regardless of frequency. Some band-pass and band-reject VCFs may also include voltage control of the bandwidth.

The SSM2040 may be operated in any of the four basic filter modes (low-pass, high-pass, band-pass, or band-reject). Figure 8-7, for example, shows a low-pass VCF circuit built around the SSM2040. (Low-pass filters are the most commonly used in electronic music applications).

The SSM2040 is internally compensated for second-order temperature effects, but not for first-order effects. Therefore, an external thermistor should be used to ensure stability. Part of the output signal may be fed back to the input for regeneration. This creates a peak around F_c, as illustrated in Fig. 8-8. The effect can be quite pronounced.

One minor problem with the SSM2040 is that it only accepts a 1-volt peak-to-peak input signal. Most VCOs used in electronic music produce a 10-volt peak-to-peak signal. Such a signal must be attenuated by a factor of ten before being fed into the VCF. An external amplifier/buffer stage can reamplify the output signal by a factor of ten to return to the 10-volt peak-to-peak standard.

Another dedicated VCF IC is the Curtis CEM3320. This chip can also be used in any of the four basic filter modes. Among other features, this device also offers voltage control of regeneration effects.

Digital Filters

Filtering can also be accomplished with digital circuits. The MF10, shown in Fig. 8-9 is a CMOS device. Its full title is *universal dual switched capacitor filter*. This chip contains two independent general-purpose filter circuits. They are quite easy to use. Only an external clock signal source and 3 or 4 resistors are required to perform any of the standard filtering functions.

A block diagram of the MF10's internal circuitry is shown in Fig. 8-10. Each of the filter sections has three outputs. One produces a low-pass filtered signal. The second is for band-pass filtering. The third output can be configured for high-pass, band-reject (notch), or all-pass (phase shift) filtering functions. The center/cut-off frequency (F_c) may be determined by the clock frequency alone, or by both the clock frequency and external resistor ratios.

A typical application for the MF10 is illustrated in Fig. 8-11. Notice that only a minimum of external components are required for this circuit. This circuit has two outputs—low-pass, and band-pass. The band-pass output may be either inverting (phase shifted 180 degrees) or noninverting.

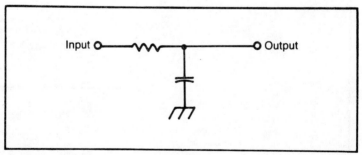

Fig. 8-6. A simple passive low-pass filter can be made up of just a resistor and a capacitor.

157

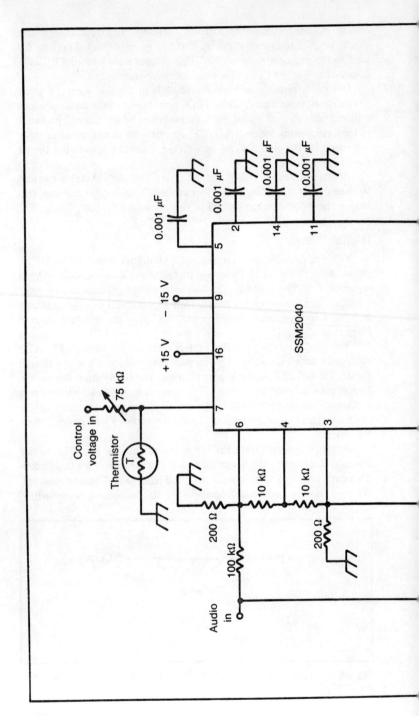

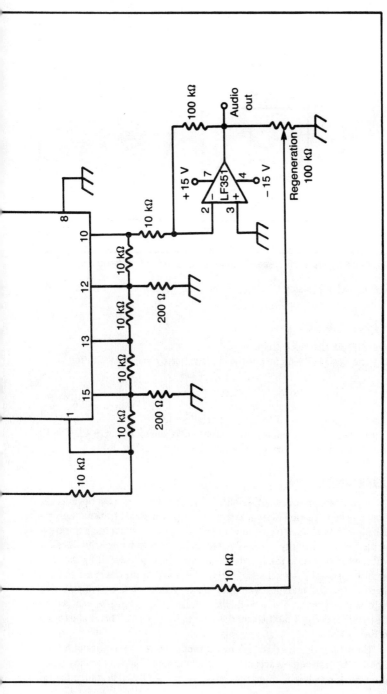

Fig. 8-7. The SSM2040 can be the heart of a high quality low-pass VCF for electronic music.

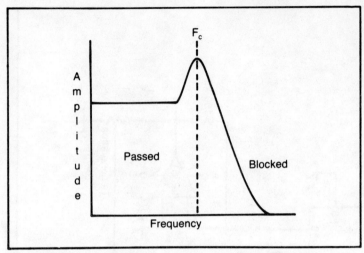

Fig. 8-8. Regeneration creates a peak around the cut-off frequency.

The center frequency (F_c) is determined by this simple formula:

$$F_c = F_{cl}/100$$

where F_{cl} is the clock frequency.

The Q (Quality factor) of the filter is set by the R3:R2 ratio:

$$Q = R3/R2$$

This circuit is intended to have a fairly low Q.

A similar circuit suitable for a higher Q is shown in Fig. 8-12. The F_c and Q equations are the same as for the circuit in Fig. 8-11.

COMPANDERS

A compander is a compression/expansion device. Its most obvious use is as a noise-reduction system in a recording, or a signal transmission application. Any storage (recording) or transmission medium has a finite dynamic range capability. If the desired signal is below a specific level, it will tend to be lost in the inevitable random noise generated by the storage/transmission system itself. On the other hand, if the desired signal exceeds a specific maximum amplitude, the excess will be clipped, and distortion will result. The recorded/transmitted signal must be kept within the dynamic range limitations of the system being used. This is illustrated in Fig. 8-13.

Unfortunately, this usable dynamic range is often not sufficient for the signal to be transmitted acceptably. Realistic reproduction of many types of music, for example, often requires a very wide dynamic range that is beyond the capabilities of economically practical recording/transmission systems.

A compander provides a solution. The signal's dynamic range is compressed at the source, then expanded for reproduction. This process is illustrated in Fig. 8-14.

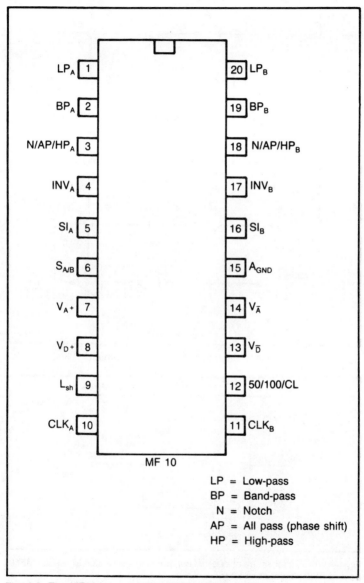

Fig. 8-9. The MF10 is a versatile digital multimode filter.

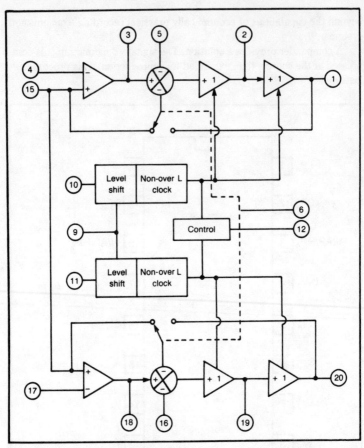

Fig. 8-10. This is a block diagram of the MF10's internal circuitry.

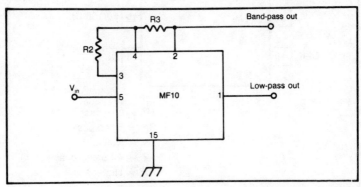

Fig. 8-11. Only a minimum of external components are needed to use the MF10 as a low-pass/band-pass filter.

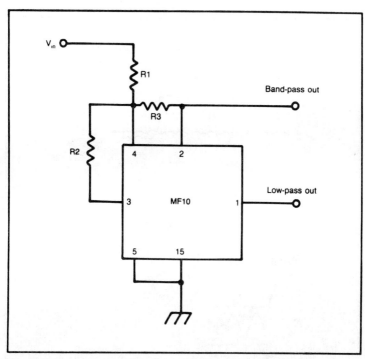

Fig. 8-12. This circuit is used when a higher Q is required than in the circuit shown in Fig. 8-11.

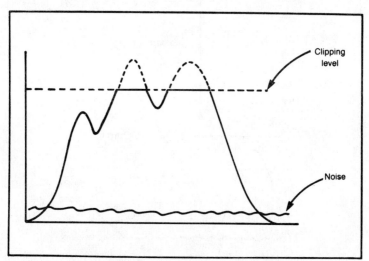

Fig. 8-13. A recorded or transmitted signal must be kept within the dynamic range limitations of the system being used.

163

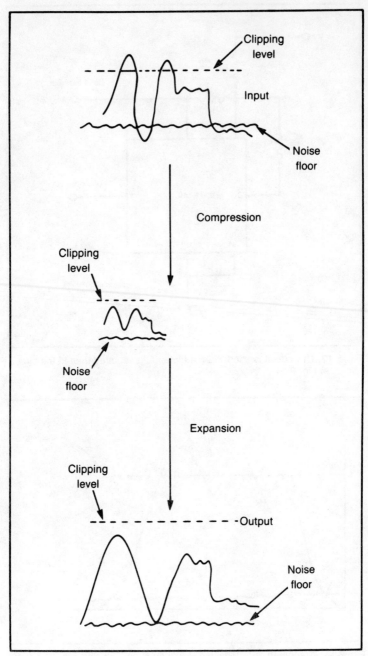

Fig. 8-14. A compander increases the total effective dynamic range of a recording or transmitting system.

The 570 Compander IC

The 570 IC is a compander circuit, designed for such applications. It contains two independent sections, which may be used for either compression or expansion. The specifications for the 570 are listed in Table 8-1, and the pinout diagram is shown in Fig. 8-15. Notice that except for pin 4 (ground) and pin 13 (Vcc), the chip is split down the middle, with the right side devoted to one complete circuit, and the left side comprising a second (identical) circuit. Except for the power-supply connections, these two sections are separate and independent. They may be used for two channels in a stereo system, or both encoding (compression) and decoding (expansion) for a single channel can be achieved with one 570 IC.

A functional block diagram of one section is shown in Fig. 8-16. The input current (from pin 2 or 15) is full-wave rectified. With an external capacitor (connected to pin 1 or 16) the signal is averaged, or smoothed out. In other words the ac signal at the input is essentially converted to a varying dc current.

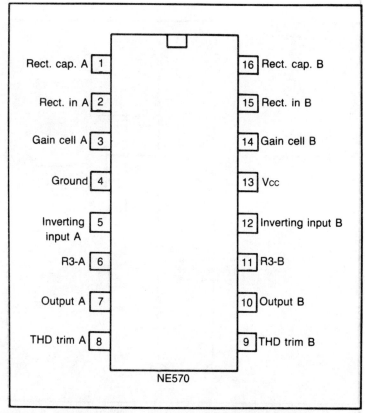

Fig. 8-15. The 570 contains two independent compander stages.

Table 8-1. This Is a Partial Specification Sheet for the 570 Compander IC.

Parameter	Minimum	Typical	Maximum	Unit
Vcc (supply voltage)	6		24	volts
Icc (supply current) (No signal)		3.2	4.0	mA
Output Current capability	±20			mA
Output slew rate		±.5		V/μS
Gain Cell Distortion (untrimmed)		.3	1.0	%
(trimmed)		.05		%
Internal reference voltage	1.7	1.8	1.9	volt
Output dc shift (untrimmed)		±20	±50	mV
Expander output noise (no signal— 20 Hz - 20 kHz)		20		μV

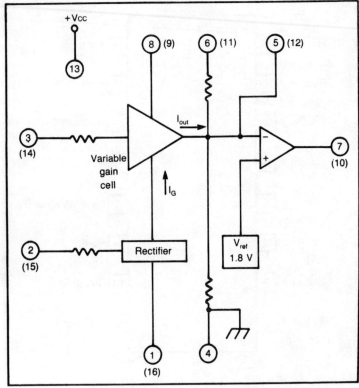

Fig. 8-16. This is a block diagram of one of the compander stages in a 570.

166

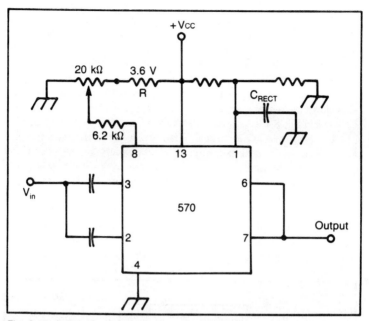

Fig. 8-17. A 570 IC can be used to create a simple compression circuit.

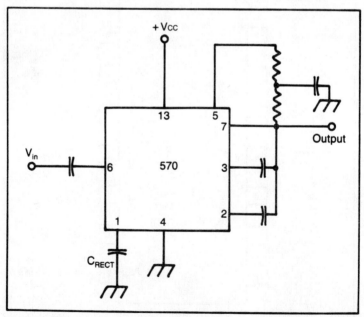

Fig. 8-18. This expansion circuit reverses the effect of the compression circuit shown in Fig. 8-17.

This rectified signal is then fed to a variable gain cell. The speed of the response to input level changes is, of course, determined by the smoothing capacitor. A trade-off is involved here. If the capacitor value is small, the gain will change rapidly, but low frequency signals may not be adequately filtered. This can result in an unpleasant warble-like effect at the output. A large value capacitor will provide better filtering, but it will also slow down the response time. A simple compression circuit using the 570 is shown in Fig. 8-17. A comparable expander circuit is shown in Fig. 8-18.

The *total harmonic distortion* (THD) can easily be trimmed at pins 8 and 9. This allows peak performance with a variety of components and circuit designs. In some cases, such as in electronic music, it may be desirable to increase the THD (and hence, the harmonic content of the output signal) by a controlled amount.

The 570 is intended for use in compander systems, but, with a little ingenuity, other applications are also possible. Because the gain of this device is determined by the signal at the rectifier input, this IC can be readily adapted for use as a VCA (*voltage-controlled amplifier*), or, with a fre-

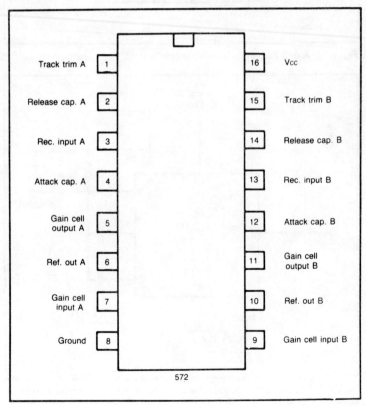

Fig. 8-19. The 572 compander is basically an improved version of the 570.

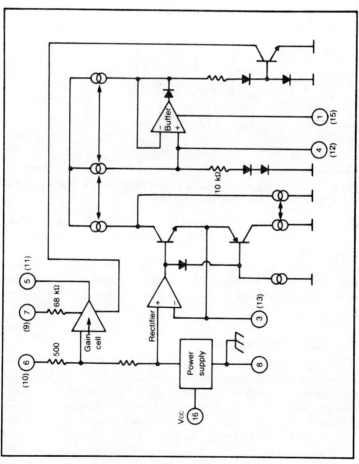

Fig. 8-20. This is a block diagram of one of the 572's independent companion channels.

169

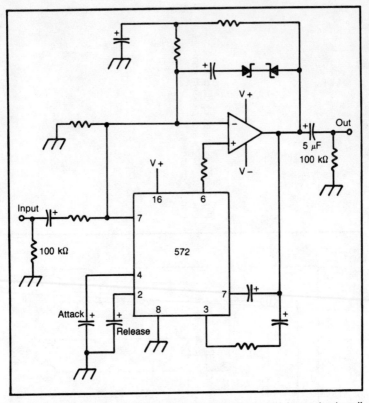

Fig. 8-21. A 572 compressor circuit is formed by placing the internal gain cell in the feedback loop of an external op amp.

quency selective network, a VCF (*voltage-controlled filter*). Another possible use is as a noise source by using a large amount of expansion, or no input signal at all.

Closely related to the 570 is the 571. The pinout is the same for both devices, and they are interchangable for most applications. The absolute maximum supply voltage for the 570 is + 24 volts, but only + 18 volts for the 571.

The 572 Compander IC

The 572 compander IC is basically an improved version of the 570. It is used primarily in noise reduction applications. Like the 570, the 572 contains two independent circuits that may be used either for compression (encoding), or expansion (decoding). The pinout diagram for the 572 is shown in Fig. 8-19. Figure 8-20 shows a block diagram of one channel.

The 572 is capable of a dynamic range of up to 109 dB, with very low distortion. Good internal filtering isolates the two channels from the com-

170

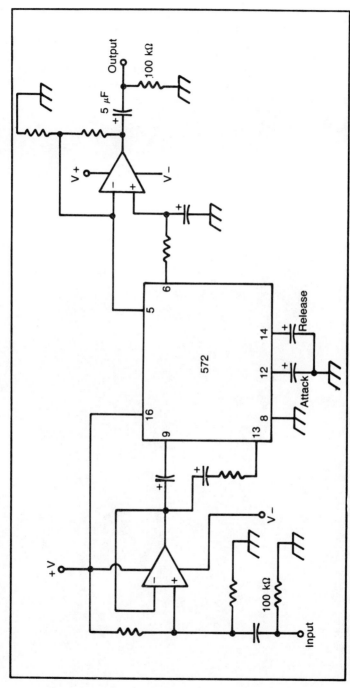

Fig. 8-22. For a 572 expander circuit, the variable gain cell is placed in series with the audio input and an external op amp.

171

mon power supply. Crosstalk between channels is so small that it can virtually be ignored.

A compressor circuit using the 572 is created by placing the internal variable gain cell in the feedback loop of an external op amp. A typical circuit is shown in Fig. 8-21. A comparable expander circuit is illustrated in Fig. 8-22. For this function the variable gain cell is placed in series with the audio input and the external op amp.

Chapter 9

Rhythm, Melody and Speech Generators

T HIS CHAPTER IS, ADMITTEDLY, SOMETHING OF A POTPOURRI. THE link between the three topics is that they all generate complex, specialized audio signals. In this chapter we will be dealing with rhythm pattern generators, melody synthesizers, and speech synthesizers.

RHYTHM PATTERN GENERATORS

Rhythm pattern generators are used to control automatic drummers. Various percussive sounds are triggered in a repeating musical pattern. Figure 9-1 shows the pinout diagram for one popular rhythm pattern generator IC. This is the MM5871, which can trigger five external sound generators in six basic rhythm patterns. Three of the rhythm patterns are in 3/4 time. They are:

- □ Slow rock
- □ Swing
- □ Waltz

The 3/4 rhythm patterns are illustrated in Fig. 9-2.

The other three rhythm patterns generated by the MM5871 are in 4/4 time, as shown in Fig. 9-3. The 4/4 patterns are:

- □ Bossa Nova
- □ Rock
- □ Samba

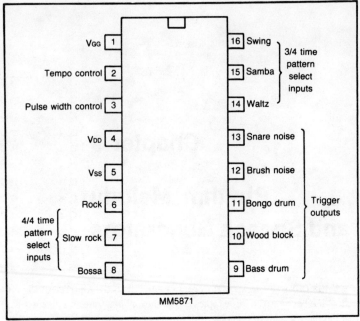

Fig. 9-1. The MM5871 is a popular rhythm pattern generator IC.

The five outputs are labelled for various standard percussion sounds. These are the instruments most drummers would use to play the rhythm patterns generated by the chip. The nominal output voices are:

☐ Bass drum
☐ Bongo
☐ Brush noise
☐ Snare noise
☐ Wood block

The circuit shown in Fig. 9-4 can be used to generate the bass drum sound. Figure 9-5 is a bongo sound generator. The wood block sound can be generated with the circuit shown in Fig. 9-6. Other percussive voices can be created with these circuits if different component values are used.

The snare and brush sounds require a random noise source. A MM5837 digital noise generator IC (see Chapter 7) does the job nicely. A snare noise circuit is shown in Fig. 9-7, and the circuit in Fig. 9-8 is designed to simulate brush noise.

In all five of these circuits, the op amps are 741's, or similar general-purpose devices. In this application there is certainly no particular need for a low-noise, or high-precision op amp.

Other sound generator circuits could also be used with the MM5871.

174

Fig. 9-2. The MM5871 can generate three rhythm patterns in 3/4 time.

3/4 Swing (Pin 16)

Output Pin	0	1	2	3	4	5
9	X					
10				X		
11	X					
12				X		
13	X			X		X

Counts

3/4 Waltz (Pin 14)

Output Pin	0	1	2	3	4	5
9	X					
10				X		
11	X					
12				X	X	
13	X			X	X	

Counts

3/4 Slow rock (Pin 15)

Output Pin	0	1	2	3	4	5
9					X	X
10	X	X	X	X	X	X
11	X			X		
12			X			
13	X	X	X	X	X	

Counts

175

4/4 Samba (Pin 6)

Output Pin	0	1	2	3	4	5	6	7
9	X	X			X			
10	X	X		X		X		
11	X	X		X				
12	X	X	X		X			
13	X	X	X	X	X	X	X	X

4/4 Rock (Pin 7)

Output Pin	0	1	2	3	4	5	6	7
9		X						
10			X	X			X	
11	X			X	X			
12			X				X	
13	X	X	X	X	X	X	X	X

4/4 Bossa (Pin 8)

Output Pin	0	1	2	3	4	5	6	7
9	X				X			
10		X		X		X		X
11							X	X
12	X		X					
13	X	X	X	X	X	X	X	X

Fig. 9-3. The MM5871 can also generate three rhythm patterns in 4/4 time.

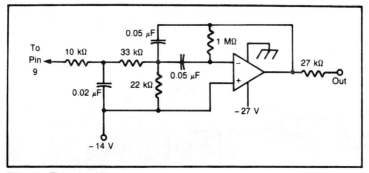

Fig. 9-4. This circuit generates a bass drum sound.

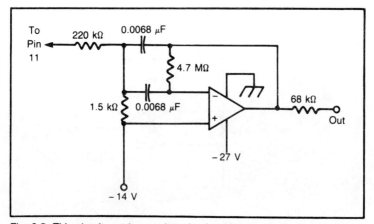

Fig. 9-5. This circuit can be used to simulate the sound of a bongo drum.

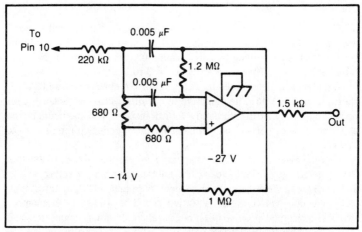

Fig. 9-6. A wood block sound can be created with this circuit.

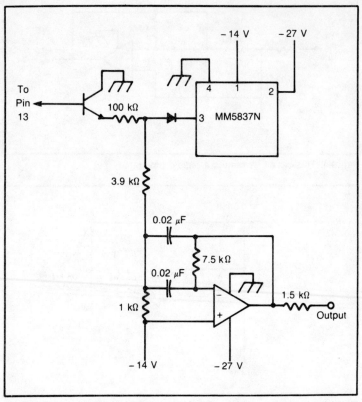

Fig. 9-7. To simulate the sound of a snare drum a noise source, like the MM5837 is used.

The MM5871 rhythm pattern generator also features variable pulse width, and an on chip tempo clock oscillator. The pulse width will typically be set for 3 ms. This is done by connecting a 0.0056 μF capacitor from pin 3 to Vss, and a 100 kΩ resistor from pin 3 to VDD.

The tempo is determined by a capacitor from pin 2 to Vss and a resistor from pin 2 to VDD. The minimum tempo of 2.7 beats per second is set up with a 0.0056 μF capacitor, and a 1.1 MΩ resistor. The maximum tempo of about 27 beats per second can be achieved by reducing the resistor to 120 kΩ.

Three power-supply connections are made to the MM5871. Vss (pin 5) is generally grounded (0 volts), and the two main supply voltages (VDD and VGG) are made negative with respect to ground. VDD (pin 4) will typically be about − 14 volts, and VGG (pin 1) will be − 27 volts. Remember that Vss is the most positive (least negative) power supply connection, and VGG is the least positive (most negative). VDD is about halfway between Vss and $VGG.

178

Figure 9-9 shows how the MM5871 is used in a typical circuit. The outputs are the sound generator circuits of Figs. 9-4 through 9-8, or similar circuits.

While only intended for the six basic rhythm patterns shown in Figs. 9-2 and 9-3, the MM5871 can be "tricked" into producing some rather unusual additional patterns, simply by grounding two or more of the pattern select pins. The results will not be as musically useful as the main patterns, but they can be rather interesting.

You can do a lot of unusual things with the MM5871. You're not limited to just automatic percussion. For example, the MM5871's outputs could be gated with outputs from a 50240 top octave generator (see Chapter 7), as shown in Fig. 9-10. This circuit will play a rhythmatic little melody.

Try using an optoisolator with a photoresistor in place of the tempo determining resistor. This is shown in Fig. 9-11. With this circuit, the tempo is voltage controllable. Many unique rhythms can be created by speeding

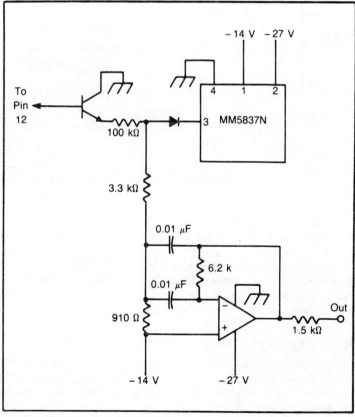

Fig. 9-8. This circuit generates a brush noise sound.

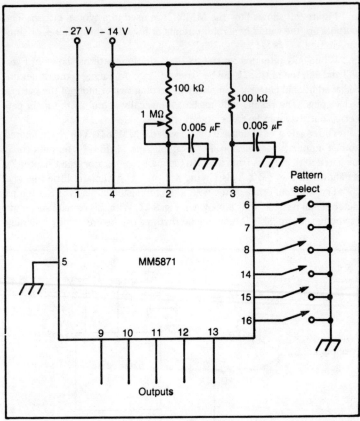

Fig. 9-9. This is a typical circuit for using the MM5871 rhythm-pattern-generator IC.

up and slowing down the tempo as the pattern is played. There is a lot you can do with a rhythm pattern generator like the MM5871, if you use a little imagination.

Another rhythm pattern generator IC is the AMI-S9660. This chip can drive seven instrument voices. The rhythm pattern is programmable, and can be up to 64 beats long. Seven rhythms can be stored for each of the seven outputs.

MELODY GENERATORS

A melody generator, or tune synthesizer is an IC that uses computer circuitry to automatically play one of several stored musical passages. They can be used in toys, doorbells, telephone ringers, car horns, music boxes, and attention-getting devices.

Epson America makes a whole line of melody generator ICs. These

180

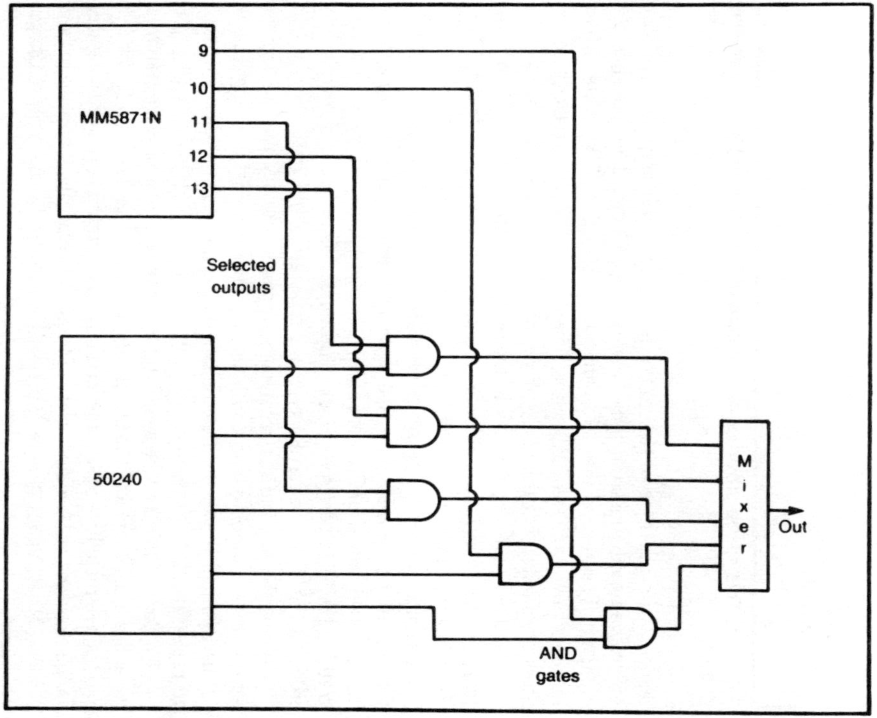

Fig. 9-10. Combining a MM5871 rhythm-pattern generator with a 50240 top-octave generator produces a simple rhythmatic melody-generator circuit.

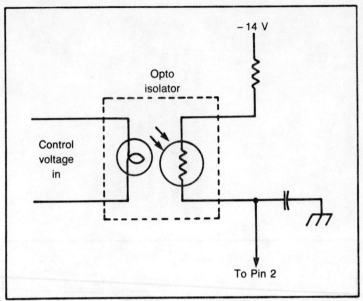

Fig. 9-11. This circuit provides voltage control over the tempo of the rhythm pattern.

devices are in three different series. The 7910 series offers two preprogrammed tunes, a chime, and an alarm tone. The melodies are diphonic. This means that two notes sound at a time. There are fifteen different devices in the 7910 series. The seven 7930 series melody generators are similar, but they only play a single tune. The third series is the 7920's. There are two devices in this series. These devices are more or less stripped down 7930's. They are contained in 8-pin DIP packages. There is no internal audio preamp/driver as in the 7930 units.

Internal circuitry processes the signal in a number of ways—note pitch, tempo, keying, tone-shaping, and other controls to generate the specific tune, chime, or alarm. For convenience in our discussion, we will concentrate on the 7910 series. A block diagram of this device is shown in Fig. 9-12. In Fig. 9-13 we have a typical circuit using the 7910.

When the appropriate terminal is brought HIGH (VDD), the melody starts at its beginning, and continues until the terminal is brought LOW. If the terminal is still HIGH at the end of the melody, the tune will start to repeat itself.

Table 9-1 indicates the switching for each of the two preprogrammed melodies, the alarm, and the chime, along with two test modes. In the test modes, the melody tempo is speeded up eight times. There are 24 devices in the three series. A wide variety of tunes are available. They range from "Silent Night" and "Jingle Bells," to a Mozart minuet and a Chopin Nocturne.

182

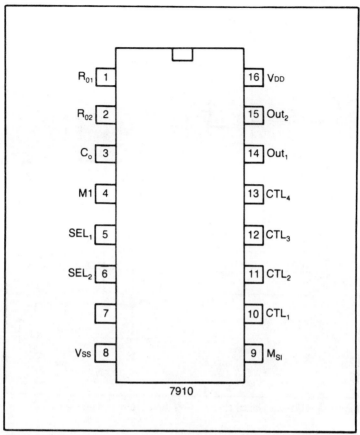

Fig. 9-12. The 7910 is an easy to use melody-generator circuit.

**Table 9-1. This Is a Summary of the
Switching to Generate the Various Tunes with the 7910.**

Function	S1	S2	S3	S4
Melody 1	1	0	0	1
Melody 2	1	0	0	1
Buzzer	x	1	0	1
Chime	x	0	1	1
Test 1	1	1	1	1
Test 2	1	1	1	1

1 = On
0 = Off
x = Don't care

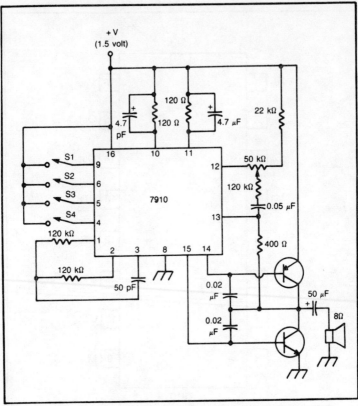

Fig. 9-13. This is a typical circuit using the 7910 melody-generator IC.

Another, even more powerful tune synthesizer is the 28-pin MP1173A from SRJ International (1936 Hillman Ave., Belmont, CA 94003). This is a full dedicated microprocessor with on chip ROM (read-only memory), RAM (random-access memory), and ALU (arithmetic logic unit). The MP1173A has thirty preprogrammed tunes. Up to sixteen melodies can be played automatically in sequence, or melodies may be manually selected with a pair of six-position rotary switches. The tunes stores in this device are based on a full chromatic scale. Each melody has its own preprogrammed tempo information. Tone, volume, and tune speed are variable.

A 50- to 60-ohm speaker can be driven directly, or an external buffer stage can be added to drive a more readily available 4- to 16-ohm speaker. This device uses very little power. Stand-by current consumption is typically about 1 μA (0.000001 amp). Maximum power consumption (assuming a 10-volt power supply) is only 100 μA (0.0001 amp). Because the current drain is so low, the device can easily be powered with batteries.

One of the most powerful tune synthesizers I am familiar with is the AY-3-1350 from General Instruments. This device's internal memory holds short segments of 25 popular melodies, plus three different chime sequences. Any of the 28 can be individually selected. The tunes programmed into the AY-3-1350 are as follows:

Toreador
William Tell
Hallelujah Chorus
Star Spangled Banner
Yankee Doodle
America, America
Deutschland Lied
Wedding March
Beethoven's Fifth
Augustine
Hell's Bells
Jingle Bells
La Vie en Rose
Star Wars
Beethoven's Ninth
John Brown's Body
Clementine
God Save The Queen
Colonel Bogey
Marseillaise
O Sole Mio
Santa Lucia
The End
Blue Danube
Brahm's Lullaby
Chime X—Westminster Chime
Chime Y—Simple Chime
Chime Z—Descending Octave Chime

And if that wasn't enough, the chip can also access an external ROM to play even more tunes. Dozens of preprogrammed EPROMS are available. Here is just a random selection of some of the tunes:

Red River Valley
O Christmas Tree
Hello Dolly
Bonanza Theme
Stranger In Paradise
Man of LaMancha
Pop Goes The Weasel
Love In Bloom

Bugle Calls
Hall Of The Mountain King
Alley Cat

and literaly hundreds more.

Only a minimum of external components are required for operation of the AY-3-1350. The pinout diagram of this 28-pin chip is shown in Fig. 9-14. The AY-3-1350 is designed for operation with a + 5 volt power supply. At the end of each tune an automatic switch-off signal is generated to avoid wasting power when the circuit is not actually in operation.

This chip has a number of interesting special features. The user is given control over the amplitude envelope for organ-like or piano-like effects. Several tunes can be automatically played in sequence. If the AY-3-1350 is to be used as a doorchime, it has capabilities for handling up to four different doors, each producing its own unique tune signal. Clearly the AY-3-1350 packs a lot of powerful circuitry into its 28-pin package.

SPEECH SYNTHESIZERS

Speech synthesis is really a digital application for the most part. It is the process of allowing a computer to produce spoken output. But since the final result, speech-like sounds, is an analog phenonemon. I am including speech synthesizers in the linear section of this book. Once again, the distinction between linear and digital devices is becoming increasingly obscure.

Consonant sounds fall into several distinct categories:

☐ Voiced consonants, such as B and L, are made up of discrete narrowband frequencies.

☐ Unvoiced consonants, such as F and S, are made up of more or less random frequencies (noise).

☐ Fricatives, such as V and Z, are basically a combination of voiced and unvoiced characteristics.

☐ Plosives, such as K, P, and T, are produced by the controlled release of a burst of air from the mouth.

A speech synthesis system must be able to simulate all four consonant types, along with the vowels. Vowels are relatively simple tone-like signals.

When a human being speaks he changes the shape of the throat, and the shape and positioning of the lips, palate, teeth, and tongue. These mouth parts, along with the respiratory system (lungs and throat) act as an adaptive filter that controls the resonances of the vocal tract. A speech synthesizer imitates these functions.

There are three primary approaches to speech synthesis in use today. They are:

☐ Direct waveform digitization

186

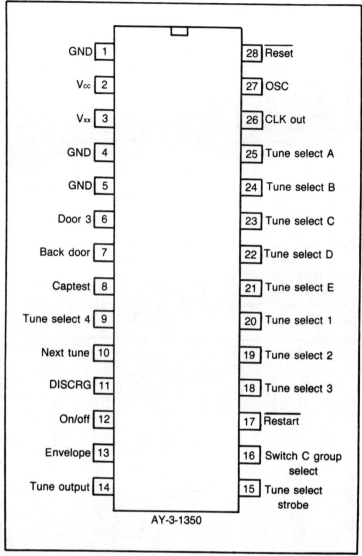

GND	1	28	$\overline{\text{Reset}}$
V$_{cc}$	2	27	OSC
V$_{xx}$	3	26	CLK out
GND	4	25	Tune select A
GND	5	24	Tune select B
Door 3	6	23	Tune select C
Back door	7	22	Tune select D
Captest	8	21	Tune select E
Tune select 4	9	20	Tune select 1
Next tune	10	19	Tune select 2
DISCRG	11	18	Tune select 3
On/off	12	17	$\overline{\text{Restart}}$
Envelope	13	16	Switch C group select
Tune output	14	15	Tune select strobe

AY-3-1350

Fig. 9-14. The AY-3-1350 is a melody-generator IC with 28 internally stored tunes.

☐ Phoneme synthesis
☐ Linear-predictive coding

We will take a brief glance at each of these methods in the following subsections of this chapter.

Direct Waveform Digitization

Conceptually, the simplest approach to speech synthesis is direct waveform digitization. In this method a live or recorded human speech signal is converted to a string of digital data by passing it through an A/D (analog-to-digital) converter. The data is stored in computer memory until needed. Then it is converted back to an analog audio signal by passing the data back through a D/A (digital-to-analog) converter.

The quality of the reproduced speech is dependent on several critical factors. The two most important factors are both limited by the amount of available memory. The sampling rate of the original speech signal is directly related to the reproduced speech quality. The more times per second the original signal is sampled and digitized, the more information will be stored about the waveshape. If an insufficient number of samples are taken, the output signal will have to approximate the missing values, resulting in distortion. Figure 9-15 shows how a one second speech signal is sampled five times, and the resulting reproduced signal. Clearly, this is not a very good reproduction of the original signal. In Fig. 9-16 the same speech signal is sampled 20 times. Note how much closer the reproduced speech signal represents the original waveform.

Now, it doesn't take much deep thinking to realize that more samples per second results in more data to be stored. This can mean a substantial increase in how much memory is required. The sampling rate must be at

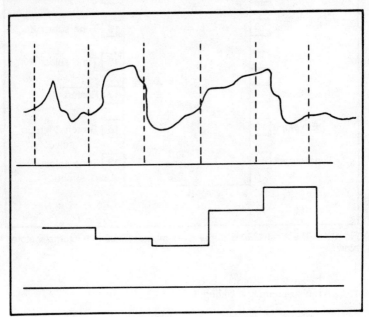

Fig. 9-15. Sampling a speech signal 5 times a second does not give very good reproduction.

188

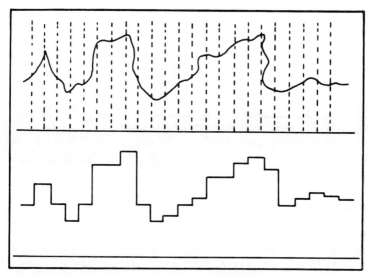

Fig. 9-16. More frequent sampling gives better fidelity.

least twice the highest input frequency of interest, or phantom frequency components that were not present in the input will appear at the output. This is called aliasing, and is illustrated in Fig. 9-17. High frequencies may appear in the output as false low frequencies.

The digitization process breaks the input signal up into discrete steps for the instantaneous amplitude. The more possible steps, the less approx-

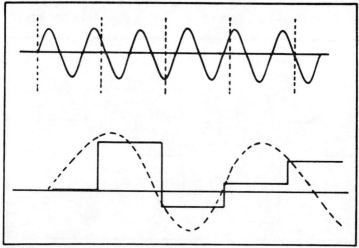

Fig. 9-17. The sampling rate must be at least twice the highest input frequency to avoid aliasing.

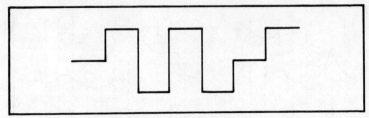

Fig. 9-18. This signal is digitized with a 3-bit system.

imation that will be required, resulting in a higher fidelity output. Unfortunately, increasing the number of amplitude steps, increases the number of bits required to store the data for each sample, and therefore the amount of memory space taken up. Figure 9-18 illustrates a signal that is digitized with a 3-bit system. There are only eight possible steps (ranging from 000 to 111). The reproduced signal is not very good. The results of using a 6-bit system are shown in Fig. 9-19. Notice that much better detail is preserved in the output signal. There is a definite trade-off in direct waveform digitization between the quality of the reproduced speech and the amount of memory required.

This is not a very memory efficient means of speech synthesis. Human speech tends to be quite redundant. For example, a drawn out sound, like "ooooh" repeats the same pattern several times. Each repetition is individually stored in the direct waveform digitization method, no matter how many times it recurs.

Special data compression techniques to recognize and reduce such redundant data are possible, and decrease the memory space consumption significantly, but the method is still the least memory efficient system of speech synthesis. On the other hand, since the generated signal is based

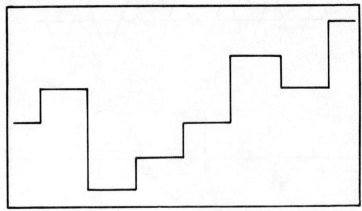

Fig. 9-19. A 6-bit digitization system gives better results than the 3-bit system illustrated in Fig. 9-18.

on actual human speech, very natural sounding speech can be created. Any voice (male, female, or child) can be reproduced, complete with intonation and accent. Pure synthesized speech often tends to sound rather unnatural.

Compressed speech data, along with frequency and amplitude data, is stored in a ROM (read-only memory), or PROM (programmable read-only memory). The speech processor module retrieves the permanently stored data, and uses it to recreate the output signal, under computer control. A block diagram of this approach is illustrated in Fig. 9-20.

The Digitalker speech synthesis system by National Semiconductor (2900 Semiconductor Ave., Santa Clara, CA 95051) uses the direct waveform digitization approach. In the Digitalker system, the compression techniques result in about 100 bits per word for a male voice. Female voices tend to require somewhat more stored data.

Three special techniques are used by the Digitalker system to compress the stored data. Redundant pitch periods are deleted. Phase adjustments and half-period zeroing cancel out the "direction" component of the signal (is it above or below the zero reference line?). This information does not affect the intelligibility of the speech in any way, so there is no need to waste memory space storing such useless data. Finally, a method called adaptive delta-modulation is used. Instead of storing the absolute amplitude value of each individual sample, only the difference between each sam-

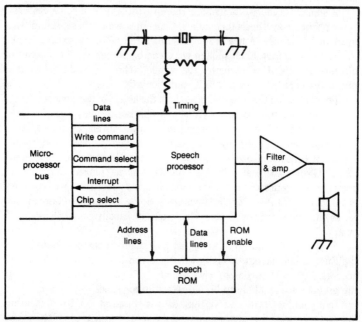

Fig. 9-20. This is a basic block diagram of the direct waveform digitization method of speech synthesis.

ple and its predecessor is recorded. These three techniques reduce the memory requirements considerably.

National Semiconductor claims that its Digitalker system is quite competitive with linear-predictive coding, which is generally considered to be the most efficient type of speech synthesis. Only a start pulse and an 8-bit address are needed to trigger any of up to 256 different messages from a single Digitalker device.

The Digitalker system is based on the MM54104 speech processor, which is shown in Fig. 9-21. A typical ROM for this system is the MS52164. It holds 8192 × 8 bits of speech data. The pinout for this ROM IC is shown in Fig. 9-2, and a listing of the stored messages is given in Table 9-2.

Phoneme Synthesis

Direct waveform digitization is certainly a straightforward method of speech synthesis, and it can achieve excellent results. But, this approach does have some significant disadvantages. It is very inefficient when it comes to memory usage. A great deal of memory is required to store the speech data, even when data compression techniques are employed. Moreover, because each word to be generated must be individually recorded and stored, a direct waveform digitization voice synthesizer always has a limited vocabulary.

A somewhat more efficient and more versatile type of voice synthesis is the phoneme synthesis techniques. Phonemes are the basic sounds that make up a language. A few dozen phonemes can be put together to make thousands of words. Each spoken word can be broken up into a series of discrete sounds. For example, the word "hello" is made up of four phonemes—"H", "short e", "L", and "long O".

Phonemes don't correspond directly to the letters of the alphabet. Many letters represent more than one phoneme. This is especially true of vowels. There are "short" vowels, as in fat, and "long" vowels, as in fate. Also, consider the difference of the "oo" sound in the words "book" and "soon".

Many consonants also have multiple sounds. For instance, there is the "hard" G sound in "goat" and the "soft" G sound in "judge." Certain letter combinations represent different phonemes. "ch", for instance, is different from any of the "c" or "h" phonemes. Some letters are redundant, as far as phonemes are concerned. The letter "c" usually sounds like an "s" or a "k".

There is some question as to just how many phonemes there are in the English language, especially when you consider various dialects and accents. For most purposes, however, about 60 phonemes cover most of the grounds we're likely to be interested in for speech synthesis.

In a phoneme synthesis system, data is stored in ROM representing the various phonemes. In programming the speech output, these phonemes are strung together in various combinations to produce different words. The vocabulary is virtually unlimited. Because each sound ever used has

192

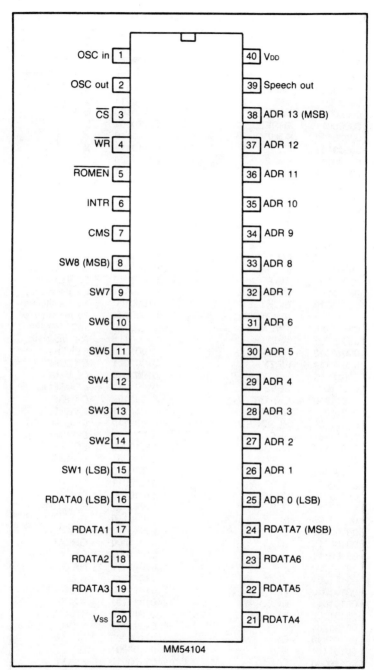

Fig. 9-21. The Digitalker system is based on the MM54104 speech processor IC.

193

Table 9-2. These Are the Stored Messages Used in the Digitalker System.

Digital Address	Message	Digital Address	Message
00000000	THIS IS DIGITALKER	00110010	S
00000001	ONE	00110011	T
00000010	TWO	00110100	U
00000011	THREE	00110101	V
00000100	FOUR	00110110	W
00000101	FIVE	00110111	X
00000110	SIX	00111000	Y
00000111	SEVEN	00111001	Z
00001000	EIGHT	00111010	AGAIN
00001001	NINE	00111011	AMPERE
00001010	TEN	00111100	AND
00001011	ELEVEN	00111101	AT
00001100	TWELVE	00111110	CANCEL
00001101	THIRTEEN	00111111	CASE
00001110	FOURTEEN	01000000	CENT
00001111	FIFTEEN	01000001	400 Hz tone
00010000	SIXTEEN	01000010	80 Hz tone
00010001	SEVENTEEN	01000011	20 ms silence
00010010	EIGHTEEN	01000100	40 ms silence
00010011	NINETEEN	01000101	80 ms silence
00010100	TWENTY	01000110	160 ms silence
00010101	THIRTY	01000111	320 ms silence
00010110	FORTY	01001000	CENTI
00010111	FIFTY	01001001	CHECK
00011000	SIXTY	01001010	COMMA
00011001	SEVENTY	01001011	CONTROL
00011010	EIGHTY	01001100	DANGER
00011011	NINETY	01001101	DEGREE
00011100	HUNDRED	01001110	DOLLAR
00011101	THOUSAND	01001111	DOWN
00011110	MILLION	01010000	EQUAL
00011111	ZERO	01010001	ERROR
00100000	A	01010010	FEET
00100001	B	01010011	FLOW
00100010	C	01010100	FUEL
00100011	D	01010101	GALLON
00100100	E	01010110	GO
00100101	F	01010111	GRAM
00100110	G	01011000	GREAT
00100111	H	01011001	GREATER
00101000	I	01011010	HAVE
00101001	J	01011011	HIGH
00101010	K	01011100	HIGHER
00101011	L	01011101	HOUR
00101100	M	01011110	IN
00101101	N	01011111	INCHES
00101110	O	01100000	IS
00101111	P	01100001	IT
00110000	Q	01100010	KILO
00110001	R	01100011	LEFT

01100100	LESSER	01111010	POINT
01100101	LESS	01111011	POUND
01100110	LIMIT	01111100	PULSES
01100111	LOW	01111101	RATE
01101000	LOWER	01111110	RE
01101001	MARK	01111111	READY
01101010	METER	10000000	RIGHT
01101011	MILE	10000001	SS
01101100	MILLI	10000010	SECOND
01101101	MINUS	10000011	SET
01101110	MINUTE	10000100	SPACE
01101111	NEAR	10000101	SPEED
01110000	NUMBER	10000110	STAR
01110001	OF	10000111	START
01110010	OFF	10001000	STOP
01110011	ON	10001001	THAN
01110100	OUT	10001010	THE
01110101	OVER	10001011	TIME
01110110	PARENTHESIS	10001100	TRY
01110111	PERCENT	10001101	UP
01111000	PLEASE	10001110	VOLT
01111001	PLUS	10001111	WEIGHT

to be stored in memory only once, memory space is not wasted with redundant data. Phoneme synthesis is quite memory efficient.

Phoneme synthesis is far from perfect, however. Generally there is no provision for accent or inflection. The generated speech sounds very flat and mechanical. This method gives that lifeless monotone that most people think of when they imagine computer generated speech. Figure 9-23 shows a block diagram of the phoneme synthesis process.

The SC-01 from Votrax was one of the first single IC phoneme synthesizers. A 6-bit code can call up 64 different phonemes. For continuous speech, only about 70 bits (less than 9 bytes) are needed for input data. This frees up the controlling computer for other, simultaneous tasks. A list of the phonemes used in the SC-01 is given in Table 9-3. A block diagram of the SC-01 is shown in Fig. 9-24. Figure 9-25 is the pinout diagram for this device.

Another phoneme synthesis IC is the SP0256 from General Instruments (600 W. John St., Hicksville, NY 11802). A pinout diagram of this 28 pin device is shown in Fig. 9-26. The SP0256 is a software-programmable digital filter that models a human vocal tract. A 16 K ROM stores both data and programming instructions. It is expandable to 491 K of ROM directly.

An internal microcontroller manages the data flow from the ROM to the digital filter. It assembles the word strings that link speech elements together for intelligible speech. The microcontroller also controls the pitch and amplitude data necessary to excite the digital filter. A typical circuit built around the SP0256 is shown in Fig. 9-27. This circuit also uses an optional external serial speech ROM IC (the SPR-16).

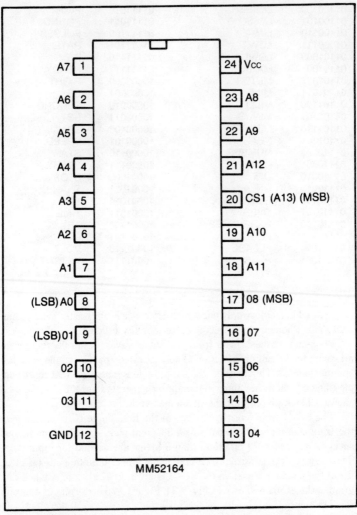

	MM52164	
A7 [1]		[24] Vcc
A6 [2]		[23] A8
A5 [3]		[22] A9
A4 [4]		[21] A12
A3 [5]		[20] CS1 (A13) (MSB)
A2 [6]		[19] A10
A1 [7]		[18] A11
(LSB) A0 [8]		[17] 08 (MSB)
(LSB) 01 [9]		[16] 07
02 [10]		[15] 06
03 [11]		[14] 05
GND [12]		[13] 04

Fig. 9-22. Speech data for the Digitalker system is stored in a MS52164 ROM IC.

Linear Predictive Coding

The third basic approach to voice synthesis is linear predictive coding. It is not quite as straightforward and obvious as the other two methods, but it offers some very definite advantages. A combination of techniques for electronically simulating the human vocal tract are used in linear predictive coding. Noise and tone sources generate signals that are processed and shaped by a special LPC filter. The next speech sample's parameters

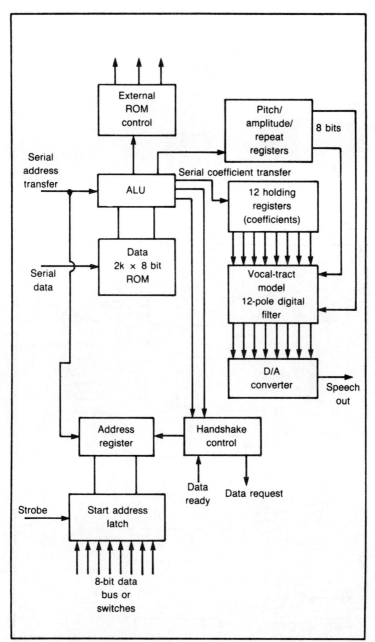

Fig. 9-23. This is a basic block diagram of the phoneme synthesis approach to speech synthesis.

197

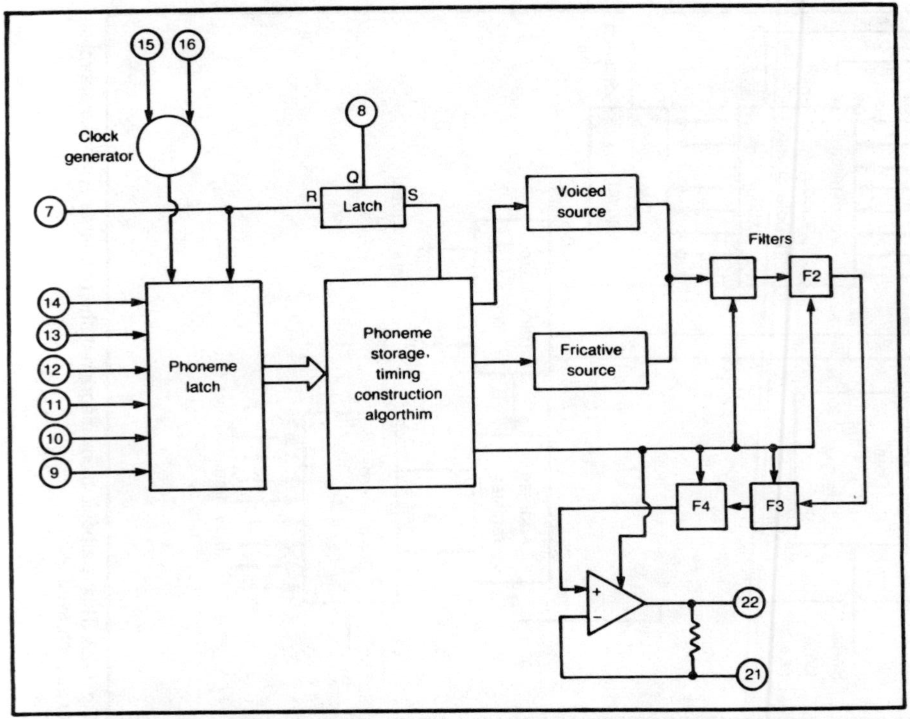

Fig. 9-24. The SC-01 is a speech synthesizer IC that uses the phoneme synthesis method.

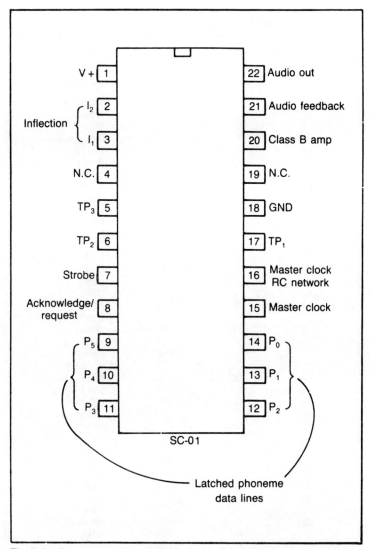

Fig. 9-25. This is the pinout diagram for the SC-01 speech synthesizer IC.

are predicted by a linear combination of the proceeding speech samples. This is clearly the source of the name. This method results in a significant reduction in the memory required to store the speech data. A block diagram of a linear predictive coding speech synthesizer is shown in Fig. 9-28. This method of speech synthesis is somewhat more complicated than the other two systems. Usually three special purpose ICs are employed. These are a digital lattice filter, a controller, and a ROM. Because of the fairly

199

Table 9-3. These Are the Phonemes
Used by the SC-01 Speech Synthesizer.

Phoneme Code	Phoneme Symbol	Duration (ms)	Sample Word
00	EH3	59	jackEt
01	EH2	71	Enlist
02	EH1	121	hEAvy
03	PA0	47	(no sound)
04	DT	47	buTTer
05	A2	71	mAde
06	A1	103	mAde
07	ZH	90	aZure
08	AH2	71	hOnest
09	I3	55	inhibIt
0A	I2	80	Inhibit
0B	I1	121	inhIbit
0C	M	103	Mat
0D	N	80	suN
0E	B	71	Bag
0F	V	71	Van
10	CH	71	CHip
11	SH	121	SHop
12	Z	71	Zoo
13	AW1	146	lAWful
14	NG	121	thING
15	AH1	146	fAther
16	OO1	103	lOOking
17	OO	185	bOOk
18	L	103	Land
19	K	80	triCK
1A	J	47	Judge
1B	H	71	Hello
1C	G	71	Get
1D	F	103	Fast
1E	D	55	paiD
1F	S	90	paSS
20	A	185	dAY
21	AY	65	dAY
22	Y1	80	Yard
23	UH3	47	missIOn
24	AH	250	mOp
25	P	103	Past
26	O	185	cOld
27	I	185	pIn
28	U	185	mOve
29	Y	103	anY
2A	T	71	Tap
2B	R	90	Red
2C	E	185	mEEt
2D	W	80	Win
2E	AE	185	dAd
2F	AE1	103	After

30	AW2	90	sAlt
31	UH2	71	About
32	UH1	103	Uncle
33	UH	185	cUp
34	O2	80	fOr
35	O3	121	abOArd
36	IU	59	yOU
37	U1	90	yOU
38	THV	80	THe
39	TH	71	THin
3A	ER	146	bIRd
3B	EH	185	gEt
3C	E1	121	bE
3D	AW	250	cAll
3E	PA1	185	(no sound)
3F	STOP	47	(no Sound)

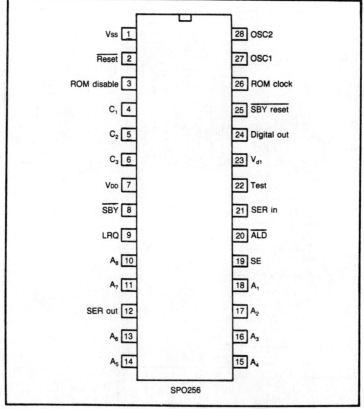

Fig. 9-26. Another phoneme synthesis based speech synthesizer IC is the SP0256.

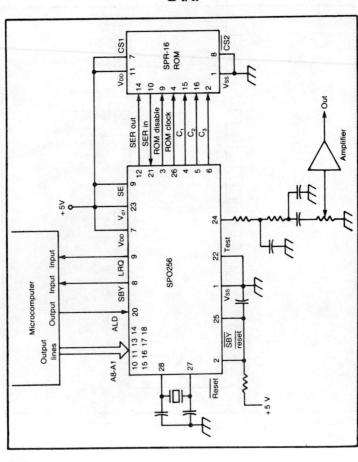

Fig. 9-27. This is a typical speech synthesizer circuit built around the SP0256.

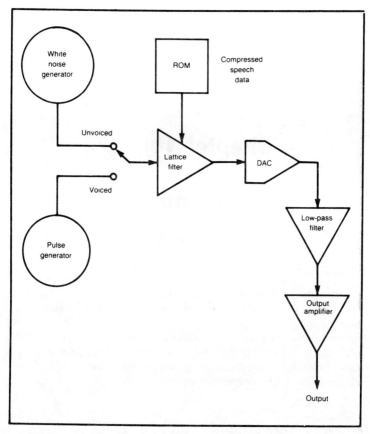

Fig. 9-28. This is a basic block diagram of the linear predictive coding approach to speech synthesis.

complex math involved in this type of speech synthesis, we will not go into detail here.

The linear predictive coding approach is used extensively by Texas Instruments (P.O. Box 225012, Dallas, TX 75265). They manufacture the TMS5100 speech processing computer, which includes an on-chip 50-mW amplifier. Another TI speech synthesizer is the TMS5220. This device was designed to be easily interfaced with an 8-bit data bus interface, making it suitable for use with most popular microcomputers. The TMS5220 was used with the TI99/4 home computer, before that product was discontinued.

```
┌──────────────────┐
│ 1   ∪        24  │
│ 2    S       23  │
│ 3    P       22  │
│ 4    E       21  │
│ 5    C       20  │
│ 6    I       19  │
│ 7    A       18  │
│ 8    L       17  │
│ 9    P       16  │
│10    U       15  │
│11    R       14  │
│12    P       13  │
│      O           │
│      S           │
│      E           │
└──────────────────┘
```

Chapter 10

Radio and
Television Video ICs

C ONSUMER ENTERTAINMENT EQUIPMENT IS A MAJOR PART OF THE
electronics industry. Special-purpose ICs for various radio and television applications have become quite common, especially with the recent
video revolution. New consumer devices are being developed every month,
and with them, new special-purpose ICs. In this chapter we will look at
a few chips that are specifically designed for use in radios, stereo systems,
television sets, and video equipment.

RF AND I-F AMPLIFIERS

In Chapter 6 we examined several amplifier ICs. These devices were
designed primarily for use with audio frequencies (up to about 20 kHz).
In a radio receiver, however, we are dealing with signals of several hundred kHz (1 kHz = 1000 Hz), or even MHz (1 MHz = 1000 kHz =
1,000,000 Hz). An audio amplifier simply can't handle such high frequencies. Either the signal will be attenuated to near nonexistence, or it will
cause the circuit to break into parasitic oscillations. Special amplifier circuits designed to handle these high frequencies are required. Of course,
several such amplifiers are available in IC form.

These high frequency amplifiers are divided into two functional classes.
The raw signal from the antenna, at the frequency being broadcast, is
boosted by an rf (*radio frequency*) amplifier. Then the signal is converted
to a somewhat lower, more convenient intermediate frequency (i-f). One
advantage of the i-f is that it is a constant frequency, across the tuner's
entire range. The rf stages have to be tuned to the desired frequency. The
i-f signal is amplified by an i-f amplifier.

The CA3002 is a general-purpose i-f amplifier. It is typically used in AM detectors and video amplifiers. A typical circuit using this device is shown in Fig. 10-1. For FM and television sound applications, a comparable device is the CA3011. The pinout diagram for this chip is shown in Fig. 10-2. The voltage gain of the CA3011 is dependent on the i-f frequency used. Three typical values are as follows:

I-f = 1 MHz   Gain = 70 dB
I-f = 4.5 MHz   Gain = 67 dB
I-f = 10.7 MHz   Gain = 61 dB

Another general-purpose i-f amplifier IC is the MC1350, shown in Fig. 10-3. This chip features wide range AGC (*automatic gain control*). The AGC range is a minimum of 60 dB from dc (0 Hz) to 45 MHz. The input and output admittance is virtually constant over the AGC range. The power gain of the MC1350 is rated at 50 dB at 45 MHz, and 48 dB at 58 MHz.

An even more versatile device is the LM273, which is shown in Fig. 10-4. A block diagram of this chip's internal circuitry is shown in Fig. 10-5. It can be used for AM, FM, or SSB (single sideband) systems. It also per-

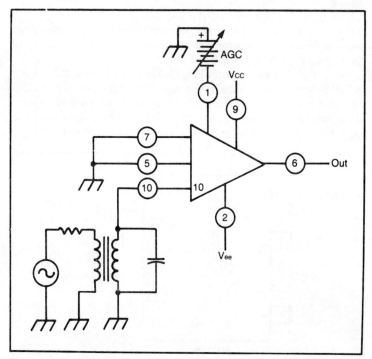

Fig. 10-1. The CA3002 general-purpose i-f amplifier is often used in AM detectors and video amplifiers.

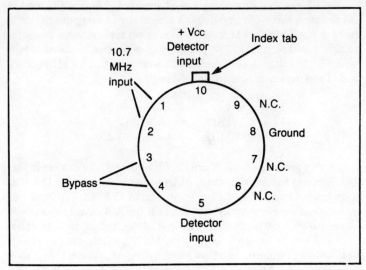

Fig. 10-2. The CA3011 is an i-f amplifier for FM and television sound applications.

forms detection. A single external filter can be used for band-pass shaping. The device may be operated at any frequency ranging from audio, all the way up to 30 MHz (30,000,000 Hz).

A closely related device is the LM373. The pinout for this chip is shown in Fig. 10-6, and the block diagram is illustrated in Fig. 10-7. The LM273 and LM373 are designed to drive low-impedance loads, such as ceramic, or mechanical filters. For high-impedance loads, such as high-Z crystal or LC filters, use the LM274 or LM374, which are otherwise similar.

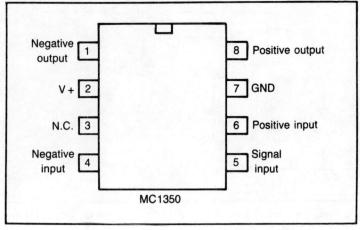

Fig. 10-3. The MC1350 is another general-purpose i-f amplifier IC.

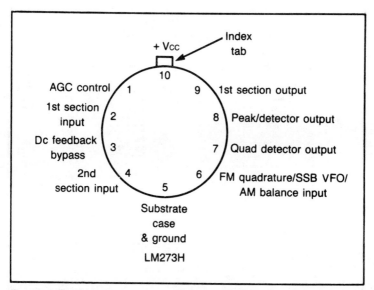

Fig. 10-4. The LM273 is a very versatile i-f amplifier IC.

Another multifunction i-f amplifier IC is the MC1358. This chip's circuitry includes an i-f amplifier, limiter, FM detector, audio driver, and electronic attenuator. In other words, this IC can cover many of the stages in a high quality FM receiver. The pinout diagram for the MC1358 is shown in Fig. 10-8.

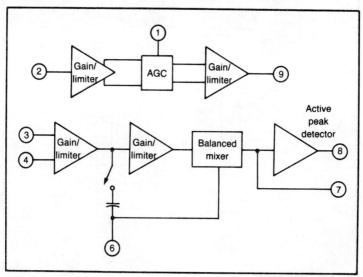

Fig. 10-5. This is a block diagram of the LM273's internal circuitry.

207

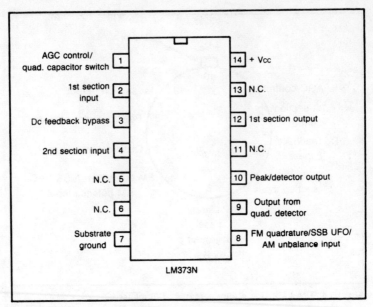

Fig. 10-6. The LM373 is closely related to the LM273.

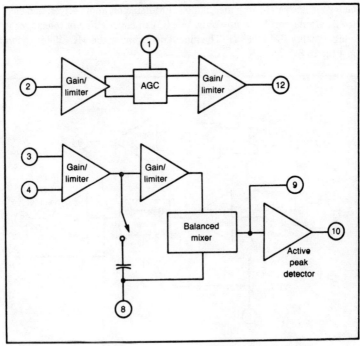

Fig. 10-7. This is a block diagram of the LM373's internal circuitry.

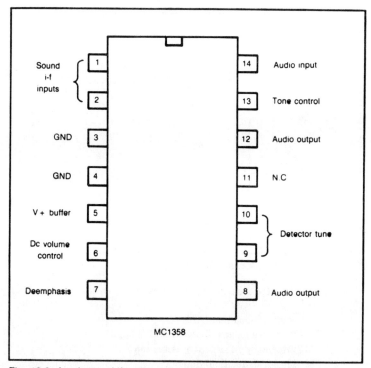

Fig. 10-8. Another multifunction i-f amplifier IC is the MC1358.

This device has some excellent specification ratings. It features high stability, low harmonic distortion, and very good AM rejection. The circuitry includes a differential peak detector that requires only a single tuning circuit. The electronic attenuator is used in place of a more conventional ac volume control. The attenuator's range is better than 60 dB. A typical rf amplifier IC is the CA3005. Its useful frequency range runs from dc (0 Hz) up to 120 MHz (120,000,000 Hz). It is packaged in a 12-lead TO-5 can. A typical circuit using the CA3005 is shown in Fig. 10-9.

OTHER RADIO CIRCUIT ICs

Of course a radio receiver includes other circuit stages in addition to the i-f and rf amplifiers. All of these stages can be obtained in IC form.

The LM1496/1596 Balanced Modulator/Demodulator

The LM1496 is a double balanced modulator/demodulator. It is available both in a round 10-pin package, as shown in Fig. 10-10, and in a 14-pin DIP housing, as illustrated in Fig. 10-11. The output voltage of this chip is proportional to the product of an input (signal) voltage, and a switching

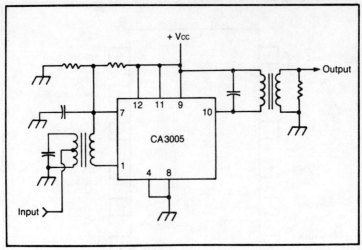

Fig. 10-9. This is typical rf amplifier circuit built around the CA3005.

(carrier) signal. There are many applications for this device. They include:

- ☐ Amplitude Modulation (AM)
- ☐ Broadband frequency doubling and chopping
- ☐ FM (Frequency Modulation) detection
- ☐ PM (Phase Modulation) detection
- ☐ Suppressed Carrier Modulation (SCM)
- ☐ Synchronous detection

The LM1496 can handle frequencies up to 100 MHz (100,000,000 Hz).

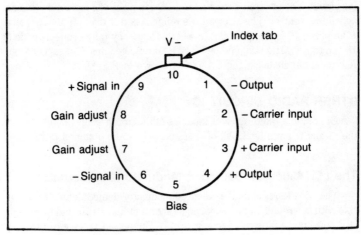

Fig. 10-10. The LM1496 is offered in a round 10-pin package.

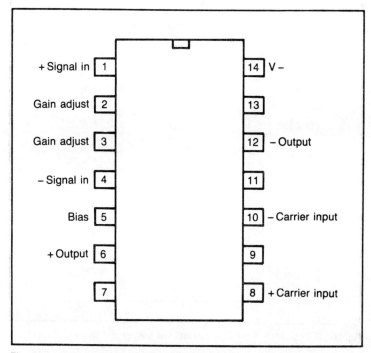

Fig. 10-11. The LM1496 is also available in a 14-pin DIP housing.

It offers excellent carrier suppression. At 10 MHz the carrier suppression is 50 dB.

The LM1496 is designed for consumer applications. It can be operated over a 0 to +70° C temperature range. A closely related device is the LM1596. The primary difference is that the LM1596 is designed for the more demanding military specifications. The operating range of temperatures for the LM1596 is –55 to +125° C. These extremes of temperature are not likely to be encountered in consumer applications.

The LM1351 FM Detector, Limiter, and Audio Amplifier

The last few stages of a simple FM receiver are contained in the LM1351, which is shown in Fig. 10-12. Only a minimum of external components are required for operation.

The LM1351 detects and demodulates the FM i-f signal, and converts it to an af signal. Three stages of i-f limiting are included, along with a balanced product detector. This chip also contains an audio preamplifier. It is capable of driving a single external class-A audio-output stage.

The LM1304/1305/1307 FM Multiplex Stereo Demodulators

Many FM radio broadcasts are in stereo. The left and right channel

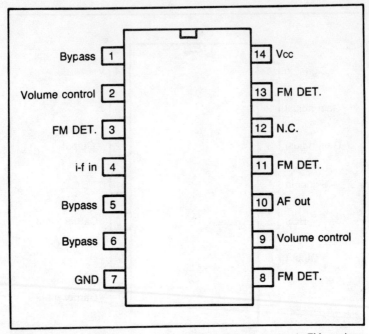

Bypass	1	14 Vcc
Volume control	2	13 FM DET.
FM DET.	3	12 N.C.
i-f in	4	11 FM DET.
Bypass	5	10 AF out
Bypass	6	9 Volume control
GND	7	8 FM DET.

Fig. 10-12. The LM1351 contains the last few stages of a simple FM receiver.

information is multiplexed as L + R and L – R signals. A stereo receiver needs a circuit to demodulate the multiplexed stereo signal. The LM1304/1305/1307 ICs are designed for this application.

These chips separate the right and left channel information for audio reproduction. No external stereo channel separation control is needed for the LM1304. The LM1305 is similar to the LM1304, but provisions are made for an external stereo channel separation control. This allows maximum channel separation.

The LM1304 also features an audio mute control, and a stereo/mono mode switch. These features are not included in the LM1307. A variation of the LM1307 is the LM1307E. This later device offers the option for emitter follower output drivers for buffers or high-current applications. A typical FM multiplex stereo demodulator circuit built around the LM1304 is illustrated in Fig. 10-13.

VIDEO AMPLIFIERS

The video signal in a television system has a wide bandwidth. Therefore the amplifier stages must be able to handle the entire width of the signal evenly. Special video amplifiers are used. Of course these circuits are available in IC form. Figure 10-14 shows a circuit using a typical video amplifier IC, the CA3001.

212

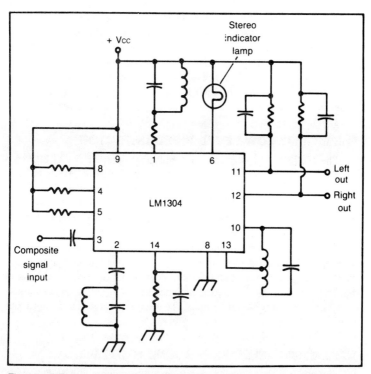

Fig. 10-13. This is a typical FM multiplex stereo demodulator circuit, using the LM1304.

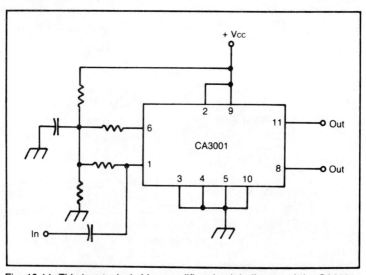

Fig. 10-14. This is a typical video amplifier circuit built around the CA3001.

213

Another video amplifier IC is the LM733. This is a two stage differential input/output wide-band video amplifier. It offers a 120 MHz (120,000,000 Hz) bandwidth. Gains of 10, 100, or 400 may be selected without any external frequency compensation components. The LM733 is available in several package types. The LM733H is in a 10-pin round can, as shown in Fig. 10-15, and the LM733D is housed in a 14-pin DIP, as illustrated in Fig. 10-16.

THE MC1330 LOW-LEVEL VIDEO DETECTOR

The MC1330 is a fully balanced multiplier detector circuit for video signals in television receivers. It is illustrated in Fig. 10-17. It can be used with either color or monochrome (black and white) television receivers. This 8-pin IC replaces several stages of a conventional TV receiver circuit, including the third i-f detector, video buffer, and AFC (automatic frequency control) buffer.

The MC1330 splits the signal into two channels. One is a straight linear amplifier, while the other channel is a limiting amplifier that provides the switching carrier for the detector. This output is buffered for use in the AFT function. This chip exhibits much less differential gain and phase distortion than many earlier circuits. The conversion gain is typically about 33 dB.

THE LM3065 TELEVISION SOUND SYSTEM

A complete television sound system can be contained within a single integrated circuit. The LM3065, which is shown in Fig. 10-18, is just such

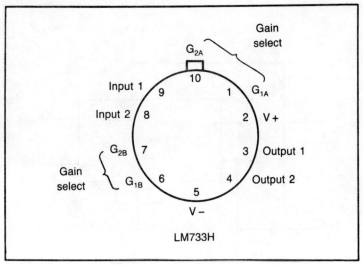

Fig. 10-15. The LM733H video amplifier comes in a 10-pin round can.

214

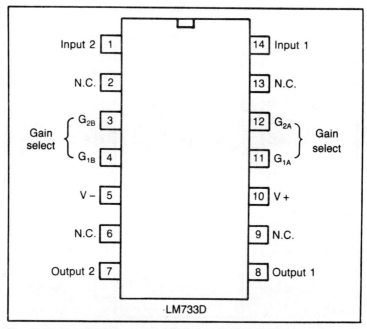

Fig. 10-16. The LM733D is available in a 14-pin DIP package.

a device. A block diagram of this chip's internal circuitry is shown in Fig. 10-19.

One interesting feature of this IC is the way the volume control is handled. An external potentiometer between pin 6 and ground varies the bias

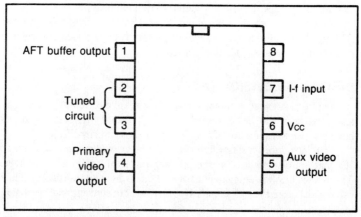

Fig. 10-17. The MC1330 is a fully balanced multiplier detector circuit for television video signals.

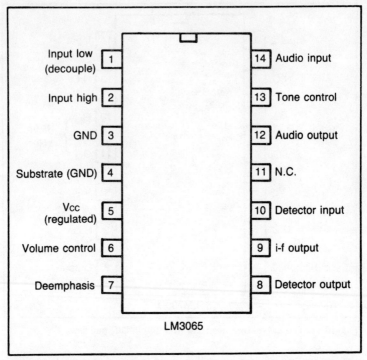

Input low (decouple)	1	14	Audio input
Input high	2	13	Tone control
GND	3	12	Audio output
Substrate (GND)	4	11	N.C.
Vcc (regulated)	5	10	Detector input
Volume control	6	9	i-f output
Deemphasis	7	8	Detector output

LM3065

Fig. 10-18. The LM3065 is a complete television sound system in a single chip.

levels of the electronic attenuator. No audio signal actually passes through the control, so hum and noise pickup are easily filtered out. Unshielded wire may be used to connect the volume control to the IC. This isn't a good idea at all for most traditional sound system circuits.

At 4.5 MHz, which is the standard i-f used for television sound, the AM rejection figure is typically 50 dB, which is quite good. The volume reduction range is greater than 60 dB.

DIGITAL TELEVISION CHIPS

Digital technology is creeping into all phases of electronics. Even many analog functions can be performed digitally. In recent years, television sets have started using digital ICs for many of their subcircuits.

ITT introduced a set of five digital VLSI (*very-large-scale integration*) ICs that are designed to for the primary stages of a television receiver. These five chips are equivalent to over 300,000 discrete transistors. They can replace more than 25 simpler ICs used in more conventional television sets.

A digital television set built around the ITT chips does not work with a new digitized signal. The regular analog signals that are being broadcast are internally converted to digital form for processing. An A/D (analog-to-digital) converter adapts the received signal to digital format. The digital signal is then processed. Finally, it is passed through a D/A (digital-to-analog) converter to be put back into analog form for the picture tube and audio amplifier.

Standard analog circuitry is still used for the first demodulation and tuning stages. To demodulate the off the air signals directly, the A/D converter would have to be capable of handling signals up to about 1 gigahertz (1,000,000 MHz). This would be prohibitively expensive and terribly impractical, at least with today's technology. In short, the "digital" television set is actually an analog/digital hybrid. A block diagram of a typical digital television receiver is illustrated in Fig. 10-20. Notice that both the input and output signals are analog signals.

The five ITT ICs shown in block diagram form in Figs. 10-21 through 10-25. The *Video Code Unit*, or VCU (Fig. 10-21) performs the A/D and

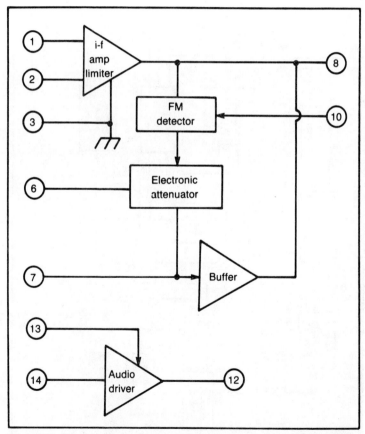

Fig. 10-19. This is a block diagram of the internal circuitry of the LM3065.

217

Fig. 10-20. Digital techniques are now being applied to television sets.

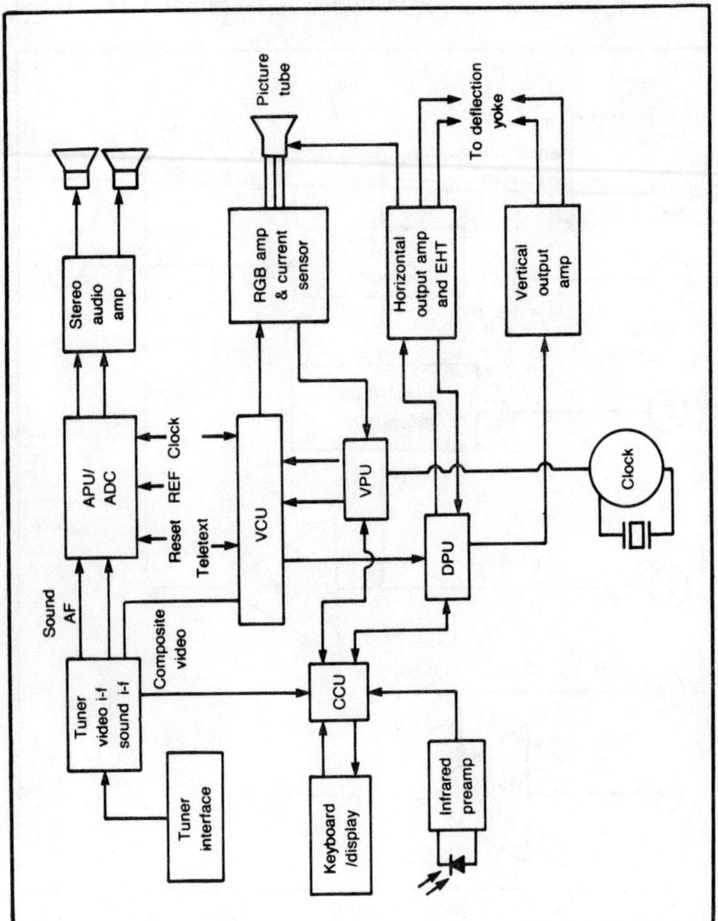

218

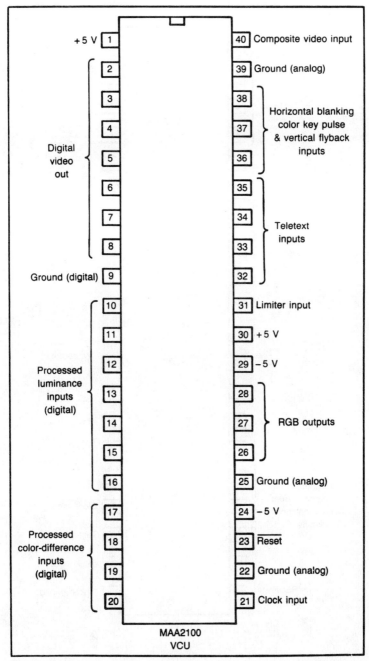

+5 V — 1 | 40 — Composite video input

2 | 39 — Ground (analog)

Digital video out — 3 | 38
4 | 37 — Horizontal blanking color key pulse & vertical flyback inputs
5 | 36

6 | 35
7 | 34 — Teletext inputs
8 | 33

Ground (digital) — 9 | 32

Processed luminance inputs (digital) — 10 | 31 — Limiter input
11 | 30 — +5 V
12 | 29 — −5 V
13 | 28
14 | 27 — RGB outputs
15 | 26
16 | 25 — Ground (analog)

Processed color-difference inputs (digital) — 17 | 24 — −5 V
18 | 23 — Reset
19 | 22 — Ground (analog)
20 | 21 — Clock input

MAA2100
VCU

Fig. 10-21. The VCU performs the A/D and D/A conversion of the signal.

219

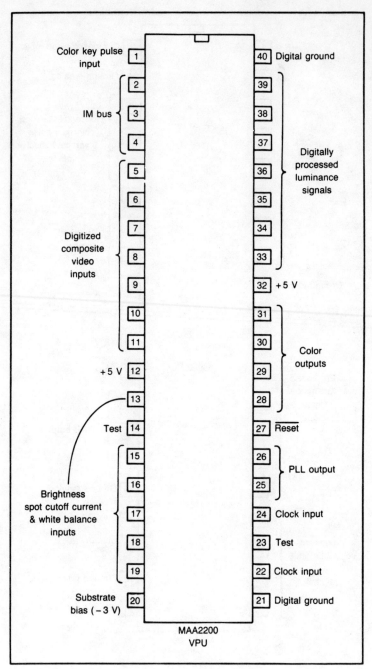

Fig. 10-22. The VPU handles the luminance and color processing.

220

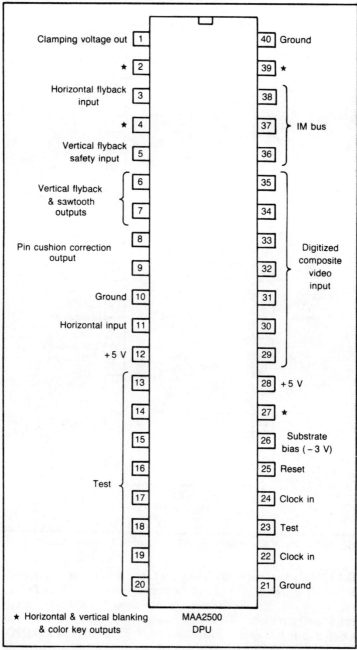

Clamping voltage out	1	40	Ground
★	2	39	★
Horizontal flyback input	3	38	⎫
★	4	37	IM bus
Vertical flyback safety input	5	36	⎭
Vertical flyback & sawtooth outputs	6	35	⎫
	7	34	
Pin cushion correction output	8	33	Digitized composite video input
	9	32	
Ground	10	31	
Horizontal input	11	30	
+ 5 V	12	29	⎭
	13	28	+ 5 V
	14	27	★
	15	26	Substrate bias (– 3 V)
	16	25	Reset
Test	17	24	Clock in
	18	23	Test
	19	22	Clock in
	20	21	Ground

★ Horizontal & vertical blanking & color key outputs

MAA2500
DPU

Fig. 10-23. Deflection processing and sync-pulse decoding are taken care of by the DPU.

221

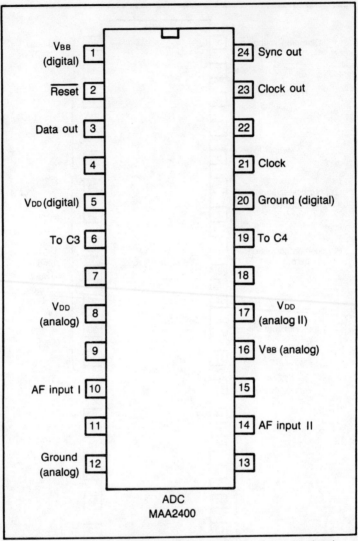

V$_{BB}$ (digital)	1	24	Sync out
$\overline{\text{Reset}}$	2	23	Clock out
Data out	3	22	
	4	21	Clock
V$_{DD}$ (digital)	5	20	Ground (digital)
To C3	6	19	To C4
	7	18	
V$_{DD}$ (analog)	8	17	V$_{DD}$ (analog II)
	9	16	V$_{BB}$ (analog)
AF input I	10	15	
	11	14	AF input II
Ground (analog)	12	13	

ADC
MAA2400

Fig. 10-24. The ADC performs the A/D conversion for the audio signal.

D/A conversion of the signal. The video signal is also split into separate R (red), G (green), and B (blue) analog output signals. The *Video Processing Unit*, or VPU (Fig. 10-22) takes care of the luminance and color processing. It can handle either NTSC (American) or PAL (European) type color processing. In Fig. 10-23, we have the *Deflection Processing Unit*, or DPU. This chip, of course, takes care of the deflection processing, and sync pulse decoding. The audio signals are handled by the *Audio Digital Converter*,

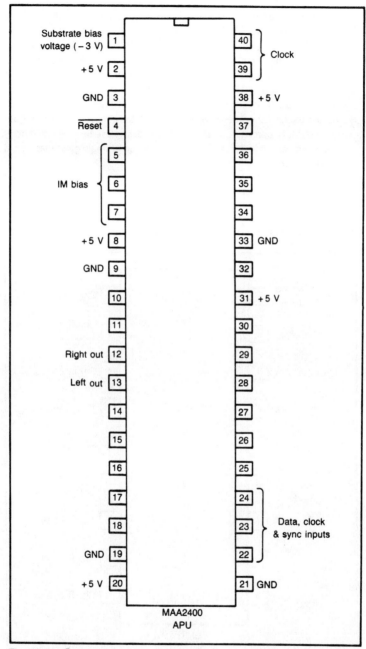

Fig. 10-25. The APU takes care of amplifier functions such as volume, tone, and balance.

223

or ADC (Fig. 10-24) and the *Audio Processing Unit*, or APU (Fig. 10-25). The ADC performs the A/D conversion for the audio signal. It can determine whether the signal is being broadcast in mono, stereo, or two-channel (bilingual) format. Amplifier functions, including volume, tone, and balance, are handled by the APU. Two independent audio channels are provided for stereo or two-channel signals.

At the time of this writing, the ITT chips are not available in small quantities to the experimenter/hobbyist, but they represent such an important new development, that they certainly deserve to be mentioned here. These, or similar devices should become available in single quantities within the next few years.

A TV MODULATOR IC

More and more video devices are being added to home television sets. These include videodiscs, videotape recorders, computers, video games,

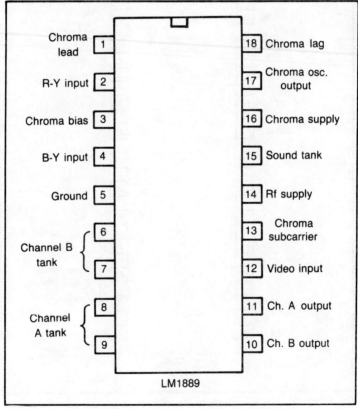

Fig. 10-26. The LM1899 is a video modulator IC for TV games, video recorders, computers, and similar devices.

and others. Many of these devices are fed into a standard television set through the antenna terminals. To do this the signal must first be converted to look to the set like a regular over-the-air broadcast signal. This is done with a device known as a TV modulator.

National Semiconductor makes a video modulator IC, known as the LM1889. The pinout diagram for this chip is shown in Fig. 10-26. The internal circuitry is illustrated in Fig. 10-27. This device generates a VHF video signal complete with color and audio.

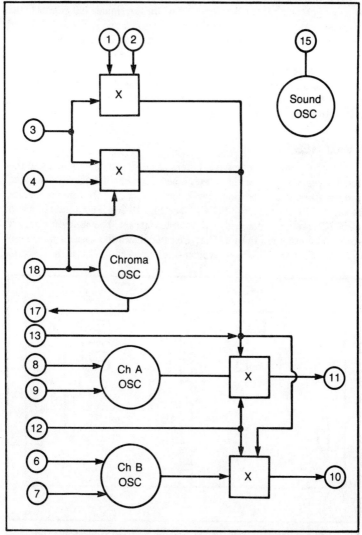

Fig. 10-27. This is a block diagram of the LM1899's internal circuitry.

The LM1889 contains two independent rf oscillators. Each may be operated at frequencies up to 100 MHz (100,000,000 Hz). In most applications these oscillators will be tuned for Channels 3 or 4. A voltage applied to the external RLC tank selects the oscillator.

A third isolated oscillator for the sound channel is also included. Its frequency can be externally frequency modulated with a varactor diode, or by switching a capacitance across the tank. The sound circuit tank components are connected between pins 15 and 16.

Still another internal oscillator is used for the color burst information. A pair of chroma modulators are fed quadrature signals (90 degrees out-of-phase with each other) from the chroma oscillator. R-Y and B-Y color difference inputs control the phase of the color output signal. A reference burst is added during the horizontal blanking period.

The LM1899 also includes on-chip voltage regulators to minimize frequency inaccuracies due to differences in the supply voltage. The rf output will be held within ± 2 kHz if the supply voltage is kept between + 12 and + 16 volts. This is good stability.

TV GAMES

Since the price of home computers has dropped so drastically in the last few years, the bottom has pretty much fallen out of the dedicated video game market. People are no longer willing to settle for the simple graphics of early video games like "PONG" (see Fig. 10-28). Thousands of video game ICs have been dumped on the surplus market. The hobbyist can pick some up for just a few dollars. They are certainly worth experimenting with.

The AY-3-8500 is a fairly typical video game chip. This is the General

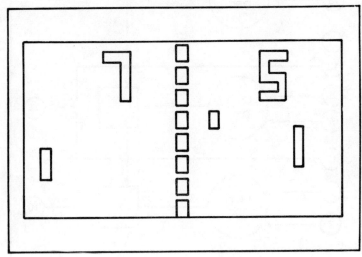

Fig. 10-28. Early video games like "PONG" were limited to very simple displays.

Instruments AY-3-8500. It is designed for six relatively simple games!

Hockey (soccer)
Practice (pelota)
Rifle shoot 1
Rifle shoot 2
Squash
Tennis (PONG)

The bat size and ball speed are externally adjustable. The ball can be fired at a variety of angles. Both automatic and manual serving are supported. Sound effects and on screen scoring (from 0 to 15) are also features of the AY-3-8500. A number of similar video game ICs are available through many electronics surplus dealers. In most cases data sheets and pinout diagrams are provided.

```
      ┌───U───┐
    1 │       │ 24
    2 │   S   │ 23
    3 │   P   │ 22
    4 │   E   │ 21
    5 │   C   │ 20
    6 │   I   │ 19
    7 │   A   │ 18
    8 │   L   │ 17
    9 │   P   │ 16
   10 │   U   │ 15
   11 │   R   │ 14
   12 │   P   │ 13
      │   O   │
      │   S   │
      │   E   │
      └───────┘
```

Chapter 11

Telephone ICs

EVER SINCE THE COURTS DECLARED THAT IT WAS NOT NECESSARILY illegal to connect non-phone company equipment to the telephone lines, there has been a flurry of telephone related equipment from a number of manufacturers. Besides actual telephones, there have been many peripheral devices, such as remote ringers, and answering machines.

Many hobbyists have also started designing and constructing their own telephone equipment. Always remember however, that any device connected to the telephone lines must be approved by the FCC (Federal Communications Commission) to be legal. Not surprisingly, this thriving new sub-industry has led to several special-purpose ICs for telephone applications.

THE MC34012 TELEPHONE RINGER IC

Sometimes we need a remote ringer for the telephone, perhaps out on the patio, or in the garage, where we might not be able to hear the phone ringing. The MC34012 is an IC designed for this type of application. It is manufactured by Motorola.

In addition, the MC34012 produces a less strident sound than the usual telephone ring. The MC34012 drives a piezoelectric ceramic sound transducer. It produces a pleasant warbling sound at a suitably high audio level. The output voltage swings over 20 volts peak-to-peak. The IC's output can source or sink up to 20 mA.

There are three different versions of the MC34012. Each is designed for a specific output frequency. Each generates two frequencies that warble (switch back and forth from one frequency to the other) at approximately a 12.5 Hz rate:

☐ The MC34012-1 has a base frequency of 4 kHz, and the output switches between 800 and 1000 Hz.

☐ The MC34012-2 has a base frequency of 8 kHz, and the output switches between 1600 Hz and 2000 Hz.

☐ The MC34012-3 has a base frequency of 2 kHz, and the output switches between 400 Hz and 500 Hz.

Except for the oscillator and output frequencies, these three devices are identical. The pinout of the MC34012 is shown in Fig. 11-1. Notice that there are no power supply connections. This chip "steals" its power from the 20 Hz ac ring-signal voltage on the telephone lines. This voltage is usually about 60 to 90 volts rms. This ac voltage is internally rectified and filtered to produce the dc voltages needed for operation of the circuit. A functional block diagram of the MC34012's internal circuitry is shown in Fig. 11-2. Only a handful of external components are required for an operating ringer circuit built around the MC34012. The basic circuit is illustrated in Fig. 11-3.

The MC34012 is connected across the telephone line through an RC series network. Resistor R1 has a value between 2 kΩ (2000 ohms) and 10 kΩ (10,000 ohms). It helps limit potentially harmful current surges that result from line transients. It also influences the ring threshold voltage. This resistor also controls the ringer's input impedance as seen by the telephone line. The MC34012's impedance meets Bell and EIA requirements, if R1's value is in the range given above.

Capacitor C1 ac couples the ringer circuit to the telephone line. It also has an effect on the input impedance at low frequencies. This capacitor's value should be in the 0.4 μF to 2 μF range.

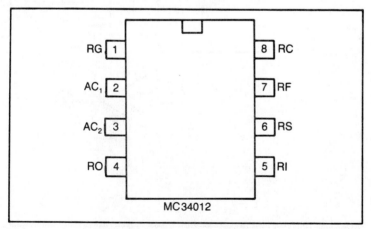

Fig. 11-1. The MC34012 replaces the bell in a telephone.

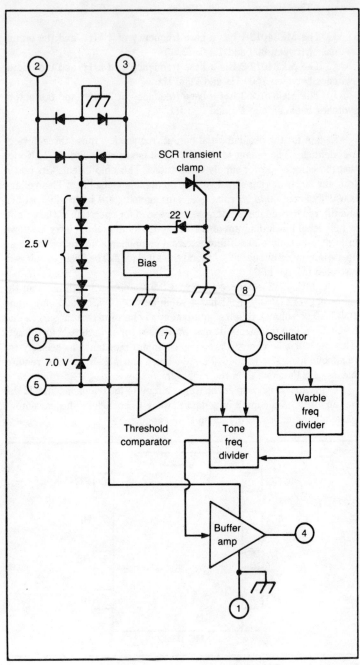

Fig. 11-2. This is a functional block diagram of the MC34012's internal circuitry.

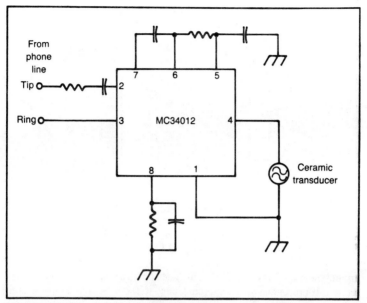

Fig. 11-3. This circuit built around the MC34012 can be used as a remote telephone ringer.

Resistor R2 and capacitor C2 control the base frequency of the oscillator. Usually these components will be selected to give the nominal base frequencies listed earlier. The values of these components should be kept within the following limits:

R2—150 kΩ (150,000 ohms) to 300 kΩ (300,000 ohms)
C2—400 pF to 2000 pF

The warble rate is internally fixed at a nominal frequency of 12.5 Hz. The output will switch back and forth between 1/5 and 1/4 of the base frequency of the oscillator.

THE TCM1512A RING DETECTOR/DRIVER

Another IC designed for remote ringers and similar applications is the TCM1512A, which is shown in Fig. 11-4. Like the MC34012, the TCM1512A adapts its operating power from the ring signal voltage itself. An on-chip full-wave regulator sets up the necessary supply voltages.

The TCM1512A's output is intended to drive a piezoelectric "sound disc" transducer. The chip outputs a warbling square wave. The two output frequencies have a 1.14:1 ratio in the 800 to 2 kHz (2000 Hz) range. The warble rate is approximately 10 Hz.

Only a few external components are required for practical circuits built

231

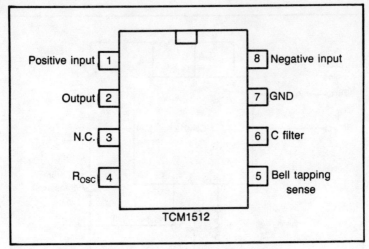

Fig. 11-4. Another telephone ringer IC is the TCM1512A.

around the TCM1512A. This chip includes built-in static and lightning protection. It presents a very high impedance to the telephone line when unactivated (stand-by). This prevents interference to any extension phones that might be in use.

Another nice feature of the TCM1512A is that dial pulses from parallel (extension) phones are ignored. This eliminates that annoying false ringing of the bell (tapping) when another phone is being dialed.

THE MC14408/MC14409 AUTOMATIC DIALER IC

Motorola also manufactures a pair of chips designed for automatic telephone dialing applications. These are the MC14408 and the MC14409. The differences between these two devices are slight. For convenience in our discussion, we will just refer to the MC14408. The same information also applies to the MC14409.

The MC14408 takes data from digital control electronics, a memory circuit, or a keypad and generates a chain of output pulses that is compatible with standard telephone circuits. This system is not related to the Touch Tone® system. Rather, the MC14408 simulates the sequential pulsing of a mechanical dialer. The pinout diagram of the MC14408 is shown in Fig. 11-5. Four-bit binary inputs are used to specify the number of output pulses to be produced:

0001	1
0010	2
0011	3
0100	4
0101	5

0110	6
0111	7
1000	8
1001	9
0000	10

The following 4-bit combinations are not used:

1010
1011
1100
1101
1110
1111

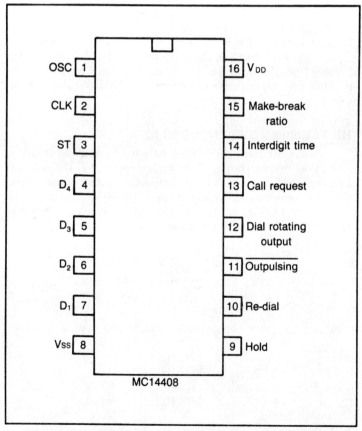

Fig. 11-5. The MC14408 is designed for automatic dialing applications.

As an example, the phone number, 875-8635 would be encoded as follows:

1000
0111
0101
1000
0110
0011
0101

An on-chip oscillator is tuned via an external LC network. The frequency of the oscillator determines the dialing rate. For example, if the oscillator is tuned to 16 kHz, ten pulses per second will be generated by the MC14408. Doubling the oscillator frequency doubles the output pulse rate.

Figure 11-6 shows a typical circuit using the MC14408, and a MC14419 2-of-8 keypad-to-binary encoder. Notice that pin 2 is the oscillator output (clock), and is tapped off to drive the MC14419. This keeps the two devices in sync with each other. The difference between the MC14408 and the MC14409 is simply what happens between each digit's pulse train. The MC14408 stays HIGH between digits, while the MC14409 goes LOW between digits.

THE TCM5089 TONE ENCODER IC

If you'd prefer to use the Touch Tone® system, you'll be interested in the TCM5089 tone encoder IC, which is shown in Fig. 11-7. This device is designed for the dual tone telephone dialing system. In this system, each digit is represented by a combination of two simultaneous tones. The controlling switches are arranged in rows and columns, as illustrated in Fig. 11-8. Each row selects one of four "low" frequencies:

697 Hz
770 Hz
852 Hz
941 Hz

At the same time, each column selects one of the four "high" frequencies:

1209 Hz
1336 Hz
1477 Hz
1633 Hz

The highest column (1633 Hz) is not used in telephone applications.

Depressing a key connects the appropriate row and column pins to

234

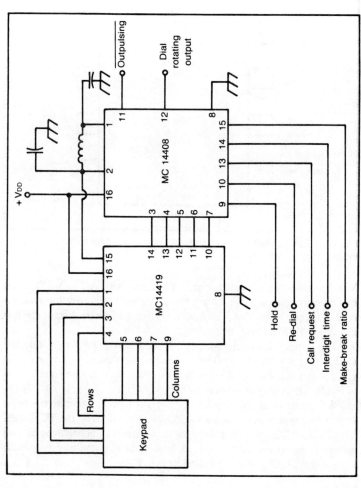

Fig. 11-6. This is a fairly typical automatic dialer circuit with keypad entry.

235

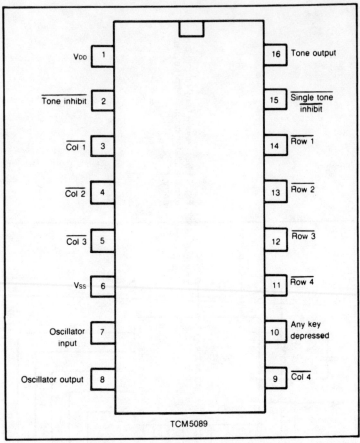

Fig. 11-7. The TCM5089 tone encoder IC can be used for applications utilizing the Touch Tone® system.

Vss. Each key corresponds to one row and one column. If more than one row or more than one column are simultaneously activated, the tone encoder's output is disabled. The output is always a combination of one low and one high frequency, or nothing.

The frequencies used may look a bit odd to you. They were selected so that they are not harmonically related to each other, or any common noise source (such as the 60 Hz house current). A valid tone combination is highly unlikely to occur accidentally.

An inexpensive TV color burst crystal is used to generate the eight sinusoidal frequencies, which are digitally synthesized by the TCM5089. The TCM5089 is often employed for data encoding applications, as well as telephone dialing applications. A basic keypad encoding circuit built around the TCM5089 is illustrated in Fig. 11-9.

236

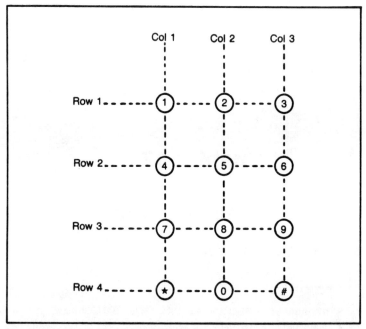

Fig. 11-8. A Touch Tone® keypad is arranged in rows and columns.

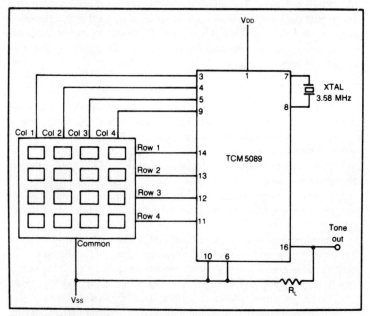

Fig. 11-9. This is a basic keypad encoding circuit using the TCM5089.

237

Chapter 12

Miscellaneous

Linear ICs

O NE PROBLEM WITH DOING A BOOK ON SPECIAL-PURPOSE IC IS THAT many devices are really special purpose. They don't fit neatly into any convenient chapter category. So we're stuck with a long chapter on miscellaneous devices. The only thing the chips in this chapter have in common is that they don't have enough in common with other ICs to belong in any of the other chapters of this book. To give this chapter some sort of organization, we will start out with relatively simple devices, and work up to some very complex chips.

THE LM339 QUAD COMPARATOR

Some special-purpose ICs are more general purpose than others. By this I mean that, while designed to perform a specific function, some of these chips can be used in a wide variety of practical applications. If I had to pick the most "general-purpose special-purpose IC," it would be a toss-up between the 555 timer (see Chapter 3) and the LM339 quad comparator, which is shown in Fig. 12-1.

A comparator does just what its name implies. It compares two voltages, and produces an output that indicates which of the input signals is larger.

A basic comparator circuit is illustrated in Fig. 12-2. A variable input signal (V_{in}) is compared to a fixed reference signal (V_{ref}). Whenever the input signal is even slightly greater than the reference signal, the output goes HIGH. Even a difference of just a few millivolts can trigger the circuit.

This simple circuit is far from perfect. If the input voltage is very close to the reference voltage, the output may tend to oscillate between states. In many applications this won't matter, but in others it could be a major

238

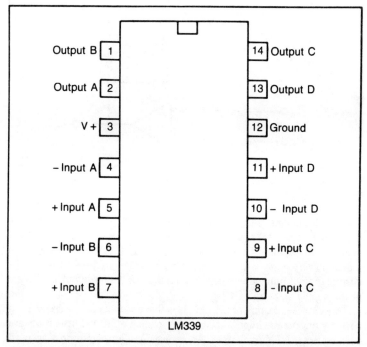

Fig. 12-1. The LM399 quad comparator is an extremely versatile special-purpose IC.

problem. One solution is to add hysteresis with a feedback resistor, as shown in Fig. 12-3. In this circuit the turn-off level is somewhat lower than the turn-on level, reducing the tendency towards input "chatter." The feedback resistor is usually given a very large value. Ten $M\Omega$ (10,000,000 ohms) is often used. The feedback resistor also speeds up the comparator's switching time, and can, in many applications, "clean up" input waveforms, functioning somewhat like a Schmitt trigger.

The comparator function can come in extremely handy in a great many circuits, including A/D (analog-to-digital) converters, detecting high/low voltage limits, driving LED displays, metering, and even monostable multivibrators. The LM339 is certainly a highly versatile special-purpose IC.

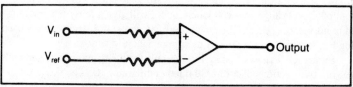

Fig. 12-2. The basic comparator circuit compares an unknown input voltage (V_{in}) with a fixed reference voltage (V_{ref}).

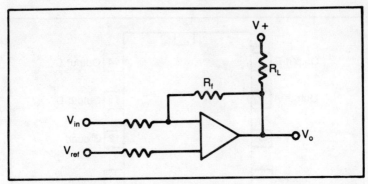

Fig. 12-3. Output "chatter" can be reduced by adding a feedback resistor to introduce hysteresis.

Besides having many applications, the LM339 is also versatile when it comes to supply voltages. It can be powered from a single polarity power supply with anything from +2 to +32 volts, or it can be operated from a dual power supply with a voltage ranging from ±1 to ±18 volts. The current drain is minimal—typically about 0.8 mA (0.0008 amp), regardless of the supply voltage.

Another nice feature of the LM339 is that its inputs can be driven from virtually any source impedance without loading effects. The outputs are open collector NPN transistors. An external pull-up resistor can be used with a different supply voltage than what is powering the rest of the chip. This makes the LM339 a good choice for interfacing to different logic families. This chip is suitable for use with any of the major logic families, including CMOS, DTL, ECL, MOS, and TTL. The comparator outputs can sink up to 20 mA (0.02 amp). Multiple outputs can be hard-wired together for an OR output.

We couldn't possibly cover all of the LM339's applications. In fact, I'm sure there are many that haven't been discovered yet. But we will glance at a few circuits using this versatile device.

Figure 12-4 shows a limit comparator. The LED will light up when the input voltage is within a specific range. If the input voltage goes below or above the specified range, the LED will be extinguished. The voltage range of interest is determined by the ratios of the resistor values in the reference voltage divider network. Since R2 is between the upper limit comparator and the lower limit comparator, its value will define the extent of the ON range. A small value for R2 will give a fairly narrow range of input voltages that will turn on the LED, while if a large value is used for R2, the turn-on range will be proportionately wider.

Figure 12-5 shows a simplified circuit for a capacitance meter built around two sections of a LM339 quad comparator IC. The unknown capacitance is linearly charged by a constant current source. The comparators are set up for a lower limit of 0.1 volt and an upper limit of 1.1 volt.

The capacitance can be determined by measuring the time it takes the capacitor to linearly charge the one volt between the low and high comparator limits.

The LM339 can also be used in monostable multivibrator applications. Figure 12-6 shows the basic LM339 monostable circuit. With no input signal the inverting input of the comparator is biased to about 1.25 volt by the voltage divider made up of R1 and R2. A negative-going trigger pulse at the input forces the inverting input below ground. Protective diode D1 limits how negative this input can go to one voltage drop.

The resulting positive-going output is fed back to the noninverting input through feedback capacitor C2. The output will stay high for a time period determined by R4 and C2. Using a 1 MΩ (1,000,000 ohms) resistor for R4 will give the following times for various values of C2:

C2	=	100 pF	T	=	0.00029 second
C2	=	1000 pF	T	=	0.0028 second
C2	=	0.01 μF	T	=	0.027 second
C2	=	0.1 μF	T	=	0.2 second

The LM339 monostable multivibrator is only suitable for relatively small time periods.

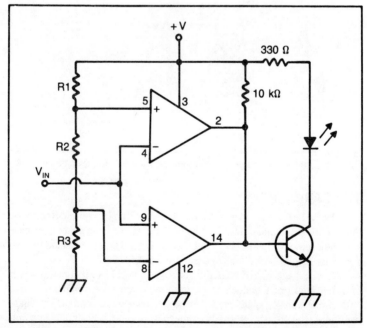

Fig. 12-4. The LED in this limit-comparator circuit will light up only when the input voltage is within a specific range.

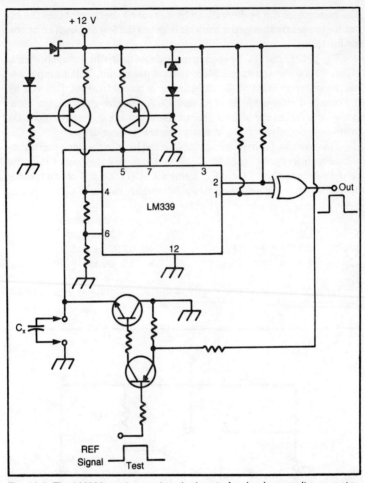

Fig. 12-5. The LM339 can be used as the heart of a simple capacitance meter.

THE GAP-01

The GAP-01, which is manufactured by Precision Monolithics Inc. (1500 Space Park Dr., Santa Clara, CA 95050), was designed as a general-purpose special-purpose IC. In fact, the name means General purpose Analog Processor. This 18-pin device's pinout is shown in Fig. 12-7.

A block diagram of the GAP-01 is given in Fig. 12-8. It contains a pair of differential transconductance amplifiers, a precision comparator, two current-mode analog switches, and an output voltage buffer amplifier. These various stages are not internally connected. Instead, their inputs and outputs are brought out to individual pins. This gives the circuit designer a great deal of versatility in arranging the stages into practical circuits.

The differential amplifiers (A and B) are connected to the output buffer under the control of the analog switches. This means the input to the buffer amplifier is digitally controllable. There are two programmable signal paths through the GAP-01.

Differential amplifier A is activated by a LOW on pin 18. A HIGH on this pin disables differential amplifier A. Differential amplifier B is controlled in just the opposite way. A LOW on pin 1 disables it, while a HIGH enables it.

The opposite logic switching allows the two control pins to be tied together, and controlled by a single digital signal. A logic LOW activates differential amplifier A, and a HIGH activates the other channel (B).

This principle is applied in the circuit shown in Fig. 12-9. This unity gain circuit can be used in many control applications. For example, it could be used to select between two microphones in an audio system. Alternatively, this circuit could select between two analog transducers in a remote telemetry system. Only one channel can ever be active at any time. One of the two channels will always be selected. The gain for each channel can be controlled by running a feedback resistor from the buffer output (pin 3) back to the appropriate input terminal.

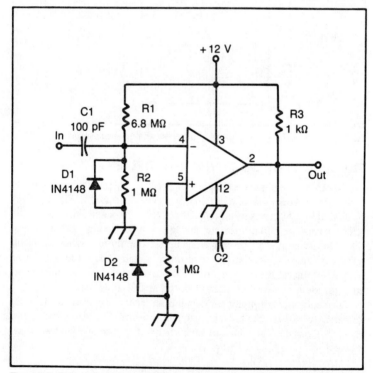

Fig. 12-6. This is the basic LM339 monostable multivibrator circuit.

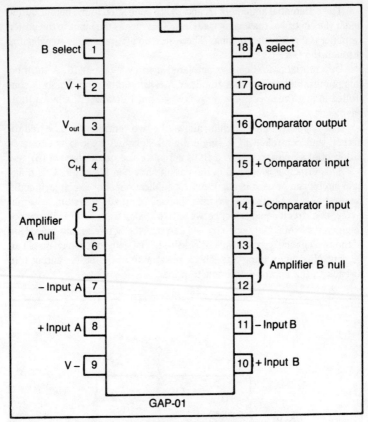

Fig. 12-7. The GAP-01 is a general-purpose analog processor.

THE XR2208 OPERATIONAL MULTIPLIER

Another special-purpose IC with a great many possible applications is the XR2208 operational multiplier, which is illustrated in Fig. 12-10. A block diagram of this chip's internal circuitry is shown in Fig. 12-11. The XR2208 contains a four-quadrant analog multiplier (modulator), an operational amplifier, and a high-frequency buffer amplifier. Notice that the op amp is not internally connected to the other circuitry (except for the power supply terminals). It can be externally wired into the circuit in any desired configuration, or it can be left out of the circuit altogether.

The output of the multiplier depends on the two input signals (X—pin 2, and Y—pin 5) and the two channel gain resistors (R_x—pins 8 and 9, and R_y—pins 6 and 7). The approximate output voltage can be found with this formula:

$$V_o = (25V_xV_y)/(R_xR_y)$$

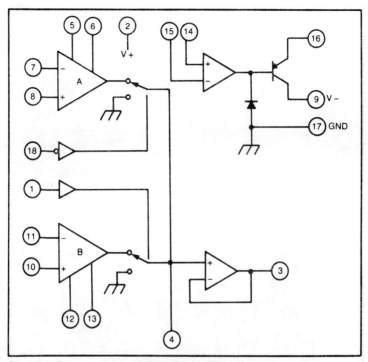

Fig. 12-8. This is a block diagram of the GAP-01's internal circuitry.

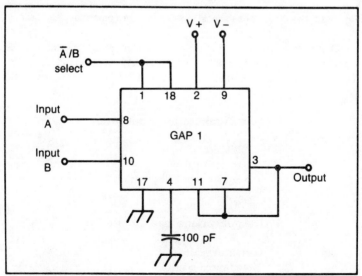

Fig. 12-9. This is a unity gain channel selector circuit.

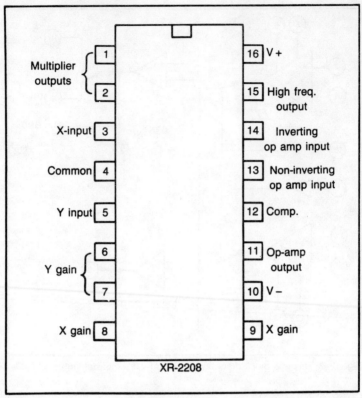

Fig. 12-10. The XR-2208 operational multiplier IC has a great many applications.

One of the multiplier outputs (pin 1) is also internally connected to the input of the buffer amplifier, which is used for high frequency applications. There are many possible applications for the XR-2208. A partial list would be as follows:

Analog Computation Applications

> Division
> Multiplication
> Square root extraction
> Squaring

PLL Applications (see Chapter 2)

> Carrier detection
> Motor speed control
> Phase-locked AM Demodulation

Signal Processing Applications

AGC amplifier
AM generation
Frequency doubling
Frequency translation
Phase detection
Synchronous AM detection
Triangle to sine-wave conversion

This chip has so many applications we can barely scratch the surface here. We will quickly glance at just a few circuits built around the XR-2208.

The circuit shown in Fig. 12-12 is a multiplication circuit. The output is directly proportional to the product of the two input signals (X and Y). Only a single input (X) is used in the circuit shown in Fig. 12-13. The output is proportional to X squared. The circuit illustrated in Fig. 12-14 performs synchronous AM detection. A triangle wave can be converted into a clean sine wave with the circuit shown in Fig. 12-15. Those are just a

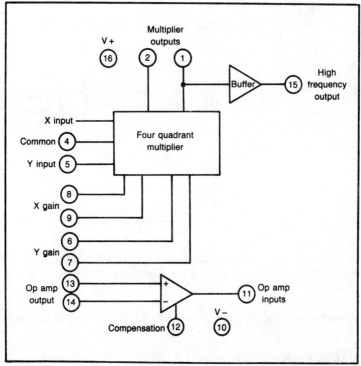

Fig. 12-11. The XR-2208 contains a four-quadrant analog multiplier, an op amp and a high-frequency buffer amplifier.

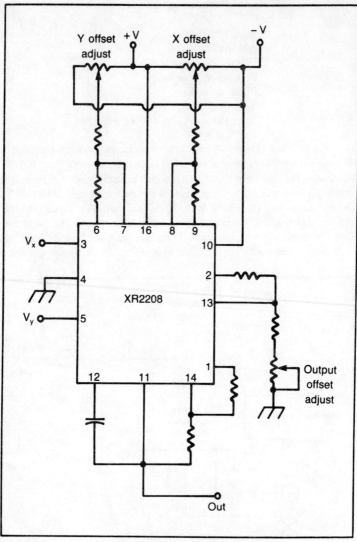

Fig. 12-12. The output of this circuit is proportional to the product of the two inputs.

few of the many possible applications for the XR2208 operational multiplier IC.

SAMPLE AND HOLD

A *sample-and-hold* is a circuit that takes an instantaneous measurement of a varying input signal and holds that value at its output until the circuit

248

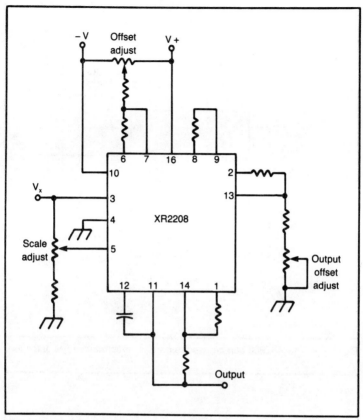

Fig. 12-13. This squaring circuit is a variation on the multiplier circuit of Fig. 12-12.

is reset. Sample-and-hold circuits (often abbreviated as S/H) are most commonly employed in measurement and A/D (analog-to-digital) conversion applications, but they have other uses too. For example, they are occasionally used in electronic music systems as a pseudorandom control voltage source.

A block diagram of a typical S/H circuit is shown in Fig. 12-16. Notice that there are two inputs to this circuit—a signal input, and a clock, or control input. As long as the clock is not triggering the circuit, the output is essentially the same as the input signal (it may be amplified or attenuated somewhat, depending on the specific circuitry used). When a clock trigger pulse is received, the circuit locks onto the instantaneous value of the input signal. This value is held frozen at the output. This output level will now be held constant, regardless of what the input signal does, until the clock is reset. The output then resumes following the input signal. This procedure is illustrated in Fig. 12-17. This description was for a theoretically ideal

249

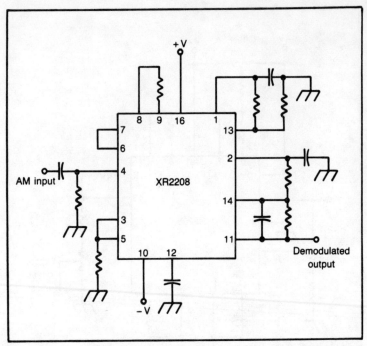

Fig. 12-14. The XR2208 can be used to perform synchronous AM detection.

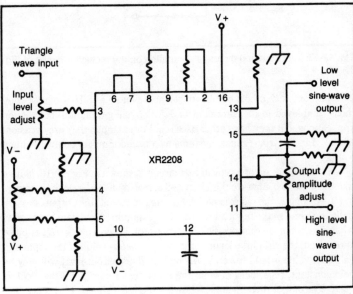

Fig. 12-15. This circuit will reshape a triangle-wave input into a sine wave output.

250

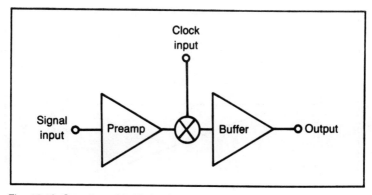

Fig. 12-16. Sample-and-hold circuits are used in many applications.

sample-and-hold circuit. Practical components force certain limitations in actual circuits. This is especially true of capacitors.

In most S/H circuits, the held output value is stored in a large capacitor. This capacitor must be capable of being charged up very quickly, while having a very low degree of leakage. That is, the tendency for the charge to drift off with time should be minimized as much as possible. No real

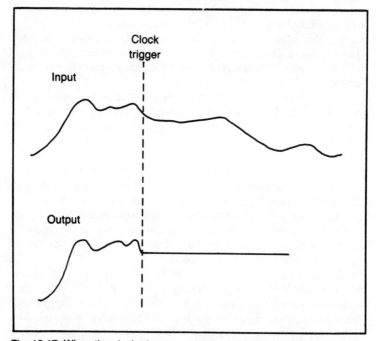

Fig. 12-17. When the clock triggers a sample-and-hold, the output "remembers" the instantaneous value of the input.

251

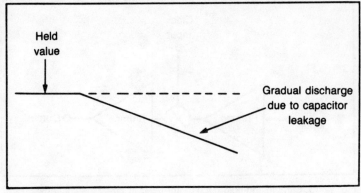

Held
value

Gradual discharge
due to capacitor
leakage

Fig. 12-18. Hold times in sample-and-hold circuits are limited by leakage of the holding capacitor.

capacitor is ideal, of course, so no sample-and-hold system can hold an output value indefinitely. Eventually the charge will leak off the storage capacitor, as illustrated in Fig. 12-18.

For the best possible results, the holding capacitor should have a low dielectric absorption figure, and extremely high insulation. These terms refer to specifications of capacitor quality. The average technician or hobbyist probably doesn't come across these terms very frequently. What they mean is that a high quality capacitor should be used. Polystyrene, mica, or tantalum capacitors are recommended. Ordinary ceramic discs, or electrolytic capacitors are not really suited for most sample-and-hold applications.

A number of sample-and-hold ICs are available. One of these devices is the HA-2420. The pinout diagram of this chip is shown in Fig. 12-19. A block diagram of the HA-2420's internal circuitry is shown in Fig. 12-20. Compare this diagram with the generalized block diagram shown in Fig. 12-16.

The input stage of the HA-2420 is a high performance op amp. This op amp was designed to offer an extremely fast slew rate, and the ability to drive highly capacitive loads without instability. This, of course, is necessary to prevent parasitic oscillations when a large-value holding capacitor is used for long holding times. Either the inverting input or the noninverting input of the op amp stage may be used separately, or both may be used together for some very unusual effects.

The real heart of any sample-and-hold circuit is the switching element, which is controlled by the clock's trigger pulses. In the HA-2420, this is a high-efficiency bipolar switching transistor stage. The leakage through this stage is extremely low when it is in its OFF condition. When the clock signal (applied to pin 14—labelled S/H Control) is HIGH, the circuit is switched into the HOLD mode. Whatever level the input signal had at the instant of switching will be maintained at the output. A LOW clock signal,

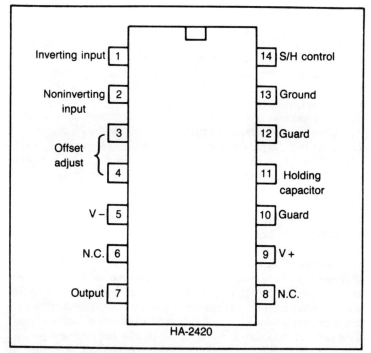

Fig. 12-19. The HA-2420 is a sample-and-hold circuit in IC form.

of course, switches the circuit into the SAMPLE mode—that is, the output simply follows the input.

Note that if the original signal is applied to the inverting input, the output will be 180 degrees out-of-phase, so when the input goes positive,

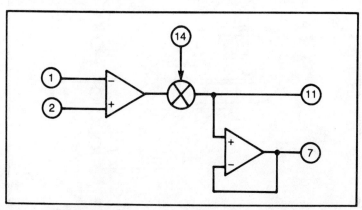

Fig. 12-20. The internal circuitry of the HA2420 sample-and-hold is quite similar to the general block diagram shown in Fig. 12-16.

the output will go negative, and vice versa.

The output stage of the HA-2420 S/H IC is another op amp, this time internally connected as a simple voltage follower, or unity gain, noninverting amplifier. This stage simply serves as a buffer for interfacing the IC with external circuitry. This helps minimize loading effects, and premature drain of the capacitor's charge. MOSFET type circuitry is used in this op amp to keep the bias current as low as possible.

To prevent excessive drift with this device, great care must be taken to minimize leakage paths in the circuit, especially on the PC board itself. Remember that a capacitor is simply two conductors separated by an insulator. Two adjacent foil traces on a PC board can easily act as a phantom capacitor. This tends to be less of a problem with audio frequencies than with rf signals. However, it is always a good idea to take a few simple precautions.

Because the output (pin 7) voltage is approximately equal to the voltage on the holding capacitor (pin 11), it can easily be employed as a guard line surrounding the capacitor connection. To facilitate this, the two pins on either side of the HOLDING CAPACITOR connection (pins 10 and 12) are designated as GUARD pins. These pins are not connected to the chip's internal circuitry. Figure 12-21 illustrates how these pins can be connected in a loop at the same potential as the output line. Any stray capacitances that occur between the PC traces will now be of little importance, because the guard line is already at the same potential as the holding capacitor.

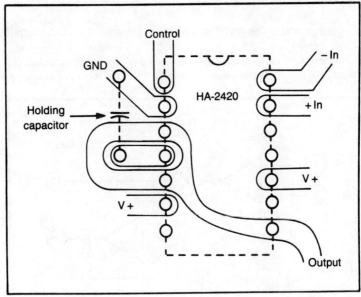

Fig. 12-21. Special guard pins on the HA2420 can help prevent leakage between adjacent foil traces on the PC board.

254

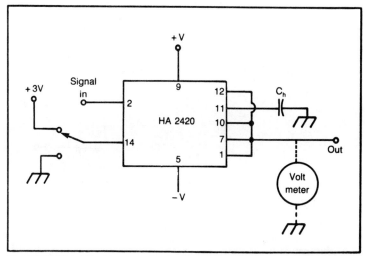

Fig. 12-22. This simple manually operated circuit demonstrates the functioning of the HA2420.

Any leakage will be irrelevant. A stray capacitance between the output line and one of the other traces on the board is not likely to matter much in most applications, thanks to the internal buffer.

Figure 12-22 shows a simple manually operated circuit using the HA-2420. This circuit is for demonstration and/or experimental purposes only. It would be of little use in the majority of practical applications.

When the switch is in its grounded position, the circuit is in the SAMPLE mode. Moving the switch to the +3 volts position places the circuit into the HOLD mode. Capacitor C, of course, is the holding capacitor. The larger the value this capacitor has, the longer the maximum hold time without excessive drift will be. However, an important trade-off is involved here. Increasing the capacitance of C reduces its charging rate, the slew rate of the S/H circuit as a whole, and the frequency bandwidth during the sample period. The means that when the circuit is switched into the HOLD mode, it may not be able to latch exactly onto the instantaneous value of the input.

A practical automatic version of this circuit may be created simply by replacing the manual control switch with a rectangle-wave oscillator (astable multivibrator), as shown in Fig. 12-23. The oscillator's signal should switch between +3 volts and ground.

It is also possible to build a sample-and-hold circuit with gain, using the HA-2420. A circuit of this type is illustrated in Fig. 12-24. The amount of gain is determined by the values of the two resistors labelled R1 and R2. The formula for the gain of this circuit is:

$$G = R1/(R1 + R2)$$

255

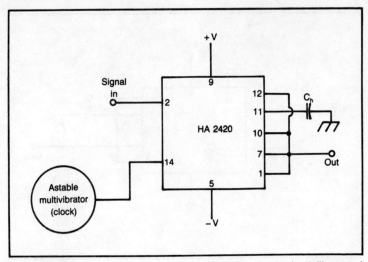

Fig. 12-23. This is a practical automatic version of the basic circuit illustrated in Fig. 12-22.

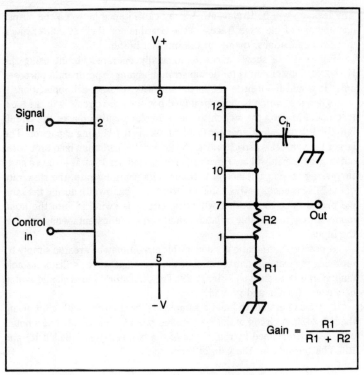

$$\text{Gain} = \frac{R1}{R1 + R2}$$

Fig. 12-24. Some sample-and-hold circuits may have gain.

Another Sample and Hold in IC form is the LH0023, which is shown in Fig. 12-25. This device can handle input signals from − 10 volts to + 10 volts, and offers up to 0.01% sample accuracy. The holding value will typically drift no more than 0.5 mV (0.0005 volt) per second. The LH0053 is another S/H IC. It is shown in Fig. 12-26. This one is designed for high-speed operation. It can acquire a 20-volt step signal in under 5 μs (0.000005 second).

FREQUENCY/VOLTAGE CONVERTERS

A frequency/voltage converter is a circuit that either converts an input frequency to a proportional output voltage, or vice versa. ICs designed for this type of function can generally be wired to operate in either direction.

In the voltage-to-frequency mode, the circuit acts essentially like a VCO with a square-wave output. As may be expected, the frequency-to-voltage mode operates in just the opposite way. The output is a continuously variable dc voltage that is directly proportional to the frequency of the signal applied to the input.

This process is extremely useful for telemetry applications. An ac signal can be easily transmitted over telephone lines, for example, but a varying dc voltage cannot. Therefore the dc signal is converted to an ac signal with a voltage-to-frequency converter. At the receiver, this process is reversed with a frequency-to-voltage converter, and the original signal is retrieved.

Figure 12-27 shows a typical frequency/voltage converter IC called the

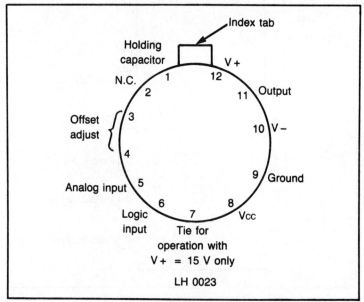

Fig. 12-25. The LH0023 is another sample-and-hold in IC form.

257

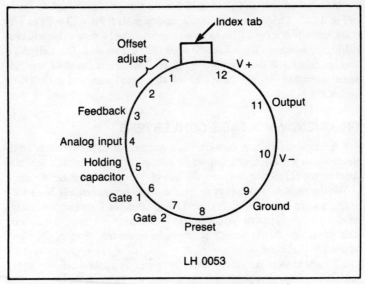

Fig. 12-26. The LH0053 is a high-speed sample-and-hold IC.

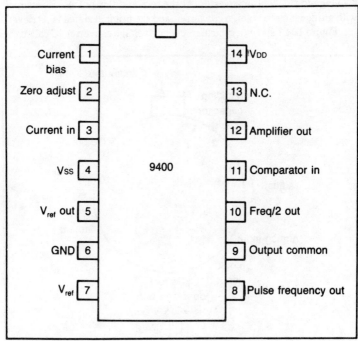

Fig. 12-27. The 9400 is frequency/voltage converter IC.

9400. This chip can be operated from a 10- to 18-volt power supply. Either a dual- or single-polarity power source may be used.

Because this device employs CMOS type circuitry, the power consumption is quite low. Typical power dissipation for the 9400 is about 27 mW (0.027 watt). Of course, like all CMOS ICs, the 9400 should be handled carefully to avoid a damaging static discharge to any of the terminal pins.

The op-amp stage is comprised of bipolar (rather than CMOS) transistors for high gain. Bipolar transistors are also used in the output stages for peak current handling. The input stages, however, use MOS type transistors to reduce offset and bias currents. Now let's examine the function of each pin of the 9400's 14-pin DIP package.

Pin 1 is the connection point for the bias resistor. The other end of this resistor should be connected to Vss (pin 4). The value of this resistor should be within the 82 kΩ to 120 kΩ range. For most applications, a 100 kΩ resistor with 10% (or better) tolerance would be a good choice.

Pin 2 is used to adjust the zero level, or low frequency set point (in the voltage-to-frequency mode). Usually a fixed voltage will be applied to this pin, although a varying control voltage could conceivably be used. This would serve as a voltage-controlled range setting. Frankly, I'm not really sure what practical advantage this might offer, but it is something that might be worth experimenting with. This pin is internally connected to the noninverting input of the operational amplifier.

Pin 3 is the current input for voltage-to-frequency conversion. This is the inverting input of the internal op amp. The nominal full-scale input current is 10 μA (0.00001 amp), but the 9400 can handle currents up to about 50 μA (0.00005 amp) without running into trouble.

If the controlling input signal is to be a voltage, rather than a current, a simple resistor can be connected in series with the control voltage source and pin 3. The value of this resistor will set the range of the circuit.

Pin 4 is the Vss, or negative voltage supply terminal.

Pin 5 is the output for the internal reference voltage. An external reference capacitor is connected between this pin and pin 3. The charging current for this capacitor is generated by the internal circuitry of the 9400, and is electronically switched.

Pin 6 is the circuit's ground connection point.

Pin 7 is the input for an external reference voltage. The accuracy of the circuit's operation will be limited by the precision of the reference voltage. This means that the reference source should be highly regulated. An IC regulator, or precision voltage source (see Chapter 1) will usually be called for, although, in some noncritical applications, you may be able to get away with a simple zener diode reference. In many applications, the reference voltage may be derived from Vss.

Pin 8 is the output connection for the voltage-to-frequency conversion mode. An open-collector bipolar transistor output stage provides a pulse wave (rectangle wave) signal, whose frequency is linearly proportional to the input current or voltage applied to pin 3. An external pull-up resistor is required for this output.

The pulse output can be directly interfaced with most major logic families, such as CMOS, DTL, MOS, and TTL. Alternately, it may be employed as an analog signal source, somewhat like a VCO (voltage-controlled oscillator).

Pin 9 is the common connection point for the outputs. The 9400's output signals are of the floating ground type, and are not intended to be referenced to circuit ground.

Pin 10 is another open-collector bipolar transistor stage. This output generates a square wave that is exactly one-half the frequency (one octave lower) than the main output signal at pin 8. In all other respects, the two outputs are identical. Once again, an external pull-up resistor is required.

Pin 11 is the input for the comparator stage. For voltage-to-frequency conversion, this pin is externally tied to pin 12 (see below). This connection triggers a 3 μs pulse delay when the input current (voltage) crosses the threshold point.

On the other hand, for frequency-to-voltage conversion, the variable frequency input signal is fed to this pin. Pin 11 is not tied to pin 12 for frequency-to-voltage conversion.

Pin 12 is the output of the internal op-amp stage. In the voltage-to-frequency mode, a descending ramp wave is available at this point. This signal has the same frequency as the signal at pin 10 (one-half the frequency at pin 8) although there is a slight phase shift. In the frequency-to-voltage mode, pin 12 produces the output voltage, which is linearly proportional to the frequency of the signal applied to pin 11.

Pin 13 is unused on the 9400, and is left disconnected. It is included only to complete the standard 14-pin DIP format.

Pin 14 is the V$_{DD}$, or positive voltage supply connection point.

Using the 9400 as a voltage-to-frequency converter is similar to using a dedicated VCO chip. The basic circuit is shown in Fig. 12-28. Notice that this circuit requires only four resistors, two capacitors, and a reference voltage source in addition to the 9400 IC itself.

With the component values shown here, this circuit can generate rectangle waves from 10 Hz to 100 kHz (100,000 Hz). Of course, the half-frequency output covers a 5 Hz to 50 kHz (50,000 Hz) range. Operation is quite linear over the circuit's entire range. Typically, the linearity is within ±0.01% up to 10 kHz.

The output varies in a direct linear manner with the input voltage. This may be somewhat problematic for electronics music applications, since most modern sound synthesis systems are designed for exponential voltage control. The problem is not insurmountable, but it must be recognized and accounted for.

Besides the input voltage, certain external component values also affect the output frequency. The complete formula for the output frequency is as follows:

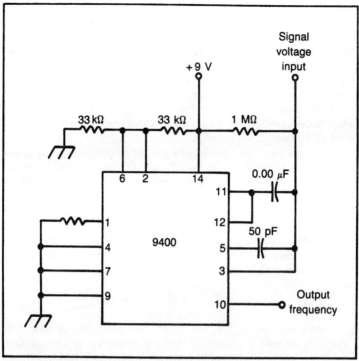

Fig. 12-28. Using the 9400 as a voltage-to-frequency converter is similar to using a dedicated VCO IC.

$$F = (V_{in}/R_{in}) \times 1/[(V_{ref})(C_{ref} + 12 \text{ pF})]$$

The input resistor will generally be selected so that the full-scale input current will be approximately 10 μA (0.00001 amp). While the exact value for the input resistor could be found by algebraically rearranging the above equation, the approximate value can be more easily obtained with this formula:

$$R_{in} = V_{inm}/10$$

where V_{inm} is the maximum input voltage. If V_{inm} is given in volts, R_{in} will be in megohms. For applications where high precision is required, a high quality multiturn trimpot can be used in place of R_{in}.

The value of the input capacitor (C_{in}) should be from three to ten times the value of C_{ref}. Best stability is achieved when C_{in} is at least four times the value of C_{ref}.

Although it does influence the output frequency, the exact value of the reference capacitor (C_{ref}) is not terribly critical. In some circuits this capacitor is made variable, and is used to trim the full scale (maximum input

261

voltage) frequency. Alternatively, a number of different capacitors could be switched in or out of the circuit for multiple ranges. The range switching could be either manual or electronic, or a combination of the two.

C_{ref} should be mounted as physically close to pins 3 and 5 as possible. Glass film, or air trimmer capacitors are preferred for maximum stability and low capacitive leakage.

In the basic frequency-to-voltage converter circuit using the 9400, a zener diode is placed between pin 6 and circuit ground sets the switching point of the internal comparator.

The input signal may be virtually any periodic waveshape, at almost any frequency from 0 Hz (dc) to 100 kHz (100,000 Hz). Up to 10 kHz, linearity within ± 0.02% can generally be expected. From 10 kHz to 100 kHz, the linearity error worsens somewhat, although linearity is fairly good throughout the 9400's entire specified range.

The input impedance of this circuit is greater than 10 megohms (10,000,000 ohms). This means loading of the signal source circuit is not likely to be much of a problem. The output voltage for a given input can be determined with this formula:

$$V_o = V_{ref} \times C_{ref} \times R_{in} \times F_{in}$$

The selected values of R_{in} and C_{ref} will also determine how rapidly the circuit can respond to changes in the input frequency. In some applications a very fast response may be required. In other applications, a slower response may be desirable. This is true in many applications in which the input signal has a lot of strong overtones or harmonics. The circuit might get confused by a strong high frequency component.

While a considerable degree of leeway is permissible in the input signal's characteristic, the internal comparator will only be properly tripped if the signal crosses through zero. Internal hysteresis of the 9400 also demands that the input signal exceed ± 200 mV (0.2 volt).

To protect the circuit, a pair of back-to-back diodes can be used to limit the peak level of high amplitude input signals. This modification will limit the input signal reaching the 9400 to ± 700 mV (0.7 volt). Of course, this will square off (clip) any waveform that exceeds this range, but since we are only concerned with the input frequency, distortion of the waveshape doesn't really matter.

While a bipolar (zero crossing) waveshape is normally called for, there are a few tricks that will allow the use of unipolar signals with the 9400. One method would be to change the reference ground point with a zener diode. Another solution would be to delete any dc component of the input signal with capacitor coupling. A third method of dealing with this problem is to employ an offset resistor network. Almost any signal source can be used to create a varying proportional voltage from the 9400.

Another fairly common voltage/frequency converter IC is the 4151. This device is sometimes numbered RC4151 or RM4151.

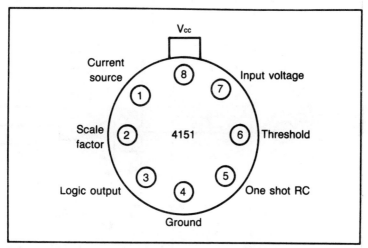

Fig. 12-29. Another fairly common frequency/voltage converter IC is the 4151.

The 4151 is a somewhat simpler chip than the 9400. It is supplied either in an 8-pin TO-99 metal can (as shown in Fig. 12-29), or an 8-pin DIP housing (Fig. 12-30).

THE TDA3810 STEREO SYNTHESIZER IC

Most people prefer to listen to music in stereo. Some music sources, however are monophonic, or have limited speaker separation (as in a car, or a portable stereo radio, in which the speakers are permanently mounted fairly close together) resulting in a reduction in the stereophonic effect.

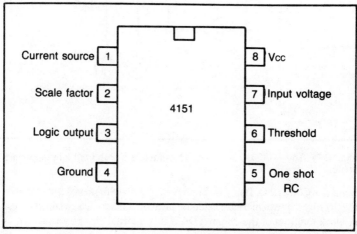

Fig. 12-30. The 4151 is also available in an 8-pin DIP housing.

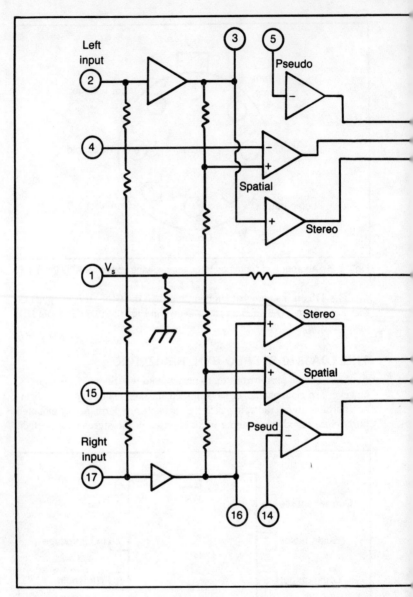

Fig. 12-31. The TDA3810 is intended to simulate stereo from a monophonic source.

Signetics Corp. (P.O. Box 409, Sunnyvale, CA 94086) makes a stereo synthesizer IC that can increase listening pleasure under such circumstances. A block diagram of the 18-pin TDA3810's internal circuitry is shown in Fig. 12-31.

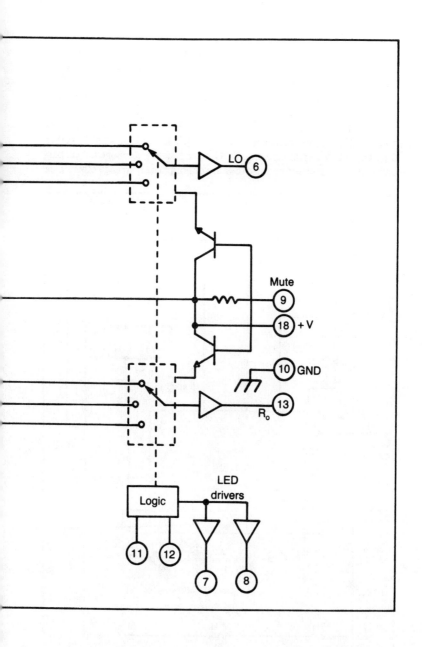

This device simulates stereo effects by using phase shifting. A monaural signal is fed to the paralleled left input (pin 2) and right input (pin 17). The signal is split into two paths. In one path, the signal is simply passed through the circuit without modification. This signal is used to drive one channel of a stereo amplifier, or a speaker. The other half of the signal, however, is partially phase shifted. All frequency components between 300 Hz and 2 kHz (2000 Hz) are delayed for a brief amount of time. The exact amount of delay varies with the frequency. For instance, 800 Hz frequency components are delayed 500 μs (0.0005 second).

This phase-shifted signal is fed to the second amplifier channel or speaker. Some of the frequency components will be in-phase with each other from the outputs, creating emphasis. Other frequency components will partially cancel each other, because they are out-of-phase with each other. This

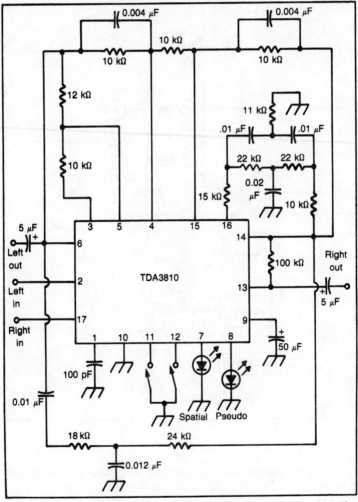

Fig. 12-32. This circuit built around the TDA3810 can operate in either the "mono" or the "spatial" mode.

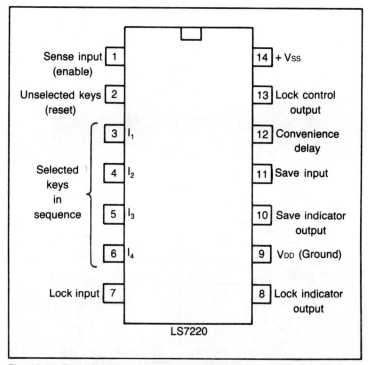

Fig. 12-33. The LS7220 is an electronic lock IC.

creates a pseudostereo effect. A similar technique is used by record manufacturers when older monophonic recordings are re-released "reprocessed for stereo."

The TDA3810 can also be employed to increase the size of the stereo image in systems where the speakers are placed close to each other, limiting the stereo effect. In this "spatial" mode, about 50% of the 180° out-of-phase feedback signal is applied as crosstalk between the right and left channels.

A typical circuit built around the TDA3810 is shown in Fig. 12-32. This circuit can operate in either the "mono" or the "spatial" mode, depending on the two switches connected to pins 11 and 12.

THE LS7220 DIGITAL LOCK IC

Security is an important area of applications. Standard mechanical locks can be picked. An electronic lock is controlled by a special circuit that will not allow the lock to be opened unless the correct code is entered. No key is required. There are usually too many possible code combinations for any-one to stumble upon the correct one by accident. In addition, many electronic lock circuits will ignore any input codes for a time after an incorrect

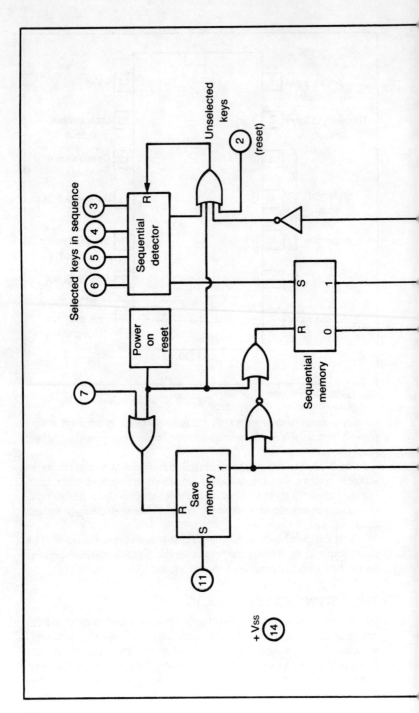

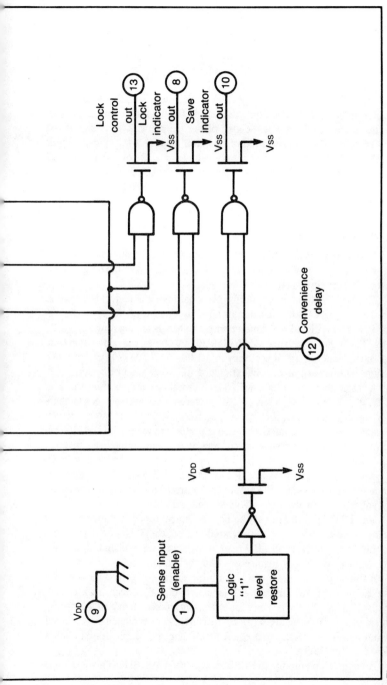

Fig. 12-34. This block diagram illustrates the functional stages of the LS7220 electronic lock.

269

entry is made. This makes experimentation very difficult, resulting in a secure lock.

The LS7220 is a special purpose IC from LSI Computer Systems, Inc. (1235 Walt Whitman Rd., Melville, NY 11747) designed for electronic lock combinations. A pinout diagram is shown in Fig. 12-33. The internal circuitry is illustrated in Fig. 12-34.

This IC features stand-alone lock logic, out-of-sequence detection, programmable convenience time delay, and 5040 possible four-digit code combinations. It uses a single-ended power supply with a voltage of +5 to +18 volts. The current drain of this chip is quite low, typically about 40 μA (0.00004 amp) when the supply voltage is +12 volts. The physical locking mechanism is controlled by a solenoid connected to the output (pin 13) of the LS7220. A typical lock circuit built around the LS7220 is shown in Fig. 12-35.

TEST EQUIPMENT ICs

A very common class of electronics circuits is test equipment for checking out other circuits and components. Several ICs have been developed for use in test equipment.

The DM7700 (manufactured by National Semiconductor) is shown in Fig. 12-36. This 24-pin chip contains all of the circuitry (except for the display itself) for a digital panel meter with a full-scale analog range of ± 1.99 volts. The input impedance is 500 kΩ (500,000 ohms), the conversion time is 1 second, and the rated accuracy is $\pm 1.0\%$. The DM7700 can produce enough output current to drive standard LED 7-segment displays.

A block diagram of the DM7700's internal circuitry is shown in Fig. 12-37. It includes an amplifier, precision reference voltage source, a voltage-to-frequency converter, a clock, a time-base counter, and latch circuits.

A/D (analog-to-digital) conversion is performed within the DM7700 using the dual voltage-to-frequency conversion technique. One V/F converter generates an ac signal with a frequency proportional to the input voltage. The second serves as a sample window, determining the clock frequency for counting the pulses in the output of the first. This IC requires two supply voltages for operation—+5 volts, and −15 volts.

The DM7700 is quite easy to use. A typical digital voltmeter circuit built around the DM7700 is illustrated in Fig. 12-38. Notice that except for the display (a NSN-33 LED readout package) only a handful of resistors and capacitors are required as external components.

A full digital voltmeter can be built from a pair of special-purpose ICs from RCA. The CA3162E is an A/D (analog-to-digital) converter with three-digit resolution. It can measure input voltage from −99 mV to +199 mV. Notice that negative voltages can be measured, even though a single-ended power supply (+5 volts) is used. A block diagram of this 16-pin chip is shown in Fig. 12-39.

A companion chip for the CA3162E is the CA3161E. This is a BCD

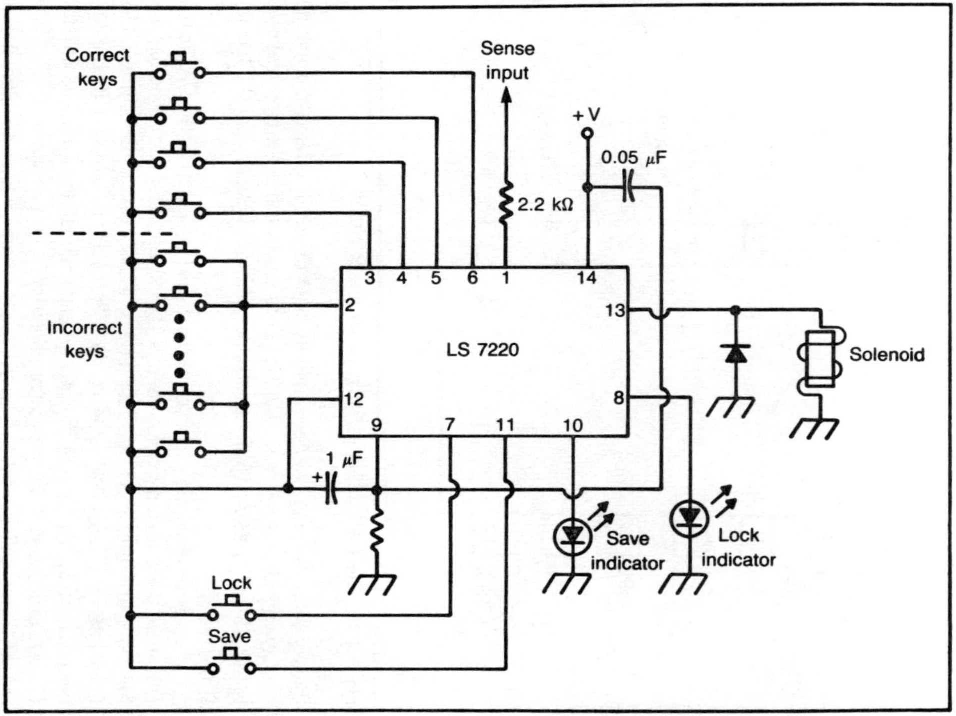

Fig. 12-35. This is a typical lock circuit built around the LS7220.

271

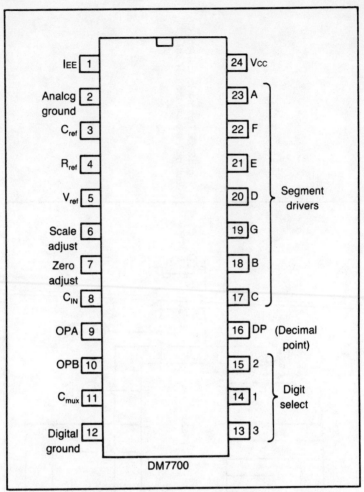

Fig. 12-36. The DM7700 is a digital panel meter circuit.

(binary-coded decimal) to 7-segment decoder/driver. It is used to drive the display with the output of the CA3162E, as illustrated in the circuit shown in Fig. 12-40. The CA3161E can also be used with virtually any TTL compatible device.

There are 16 possible logic combinations at the inputs of the CA3161E. Of course there are only ten numerals (0 to 9), but the extra six code combinations are intended to display letters, a dash, and a blank digit. The CA3162E generates 12 codes to the CA3161E. In addition to the ten numerals, a dash is used to indicate an underrange condition, and an "E" is used to represent an overrange condition.

Ac voltages and currents have always been somewhat difficult to meas-

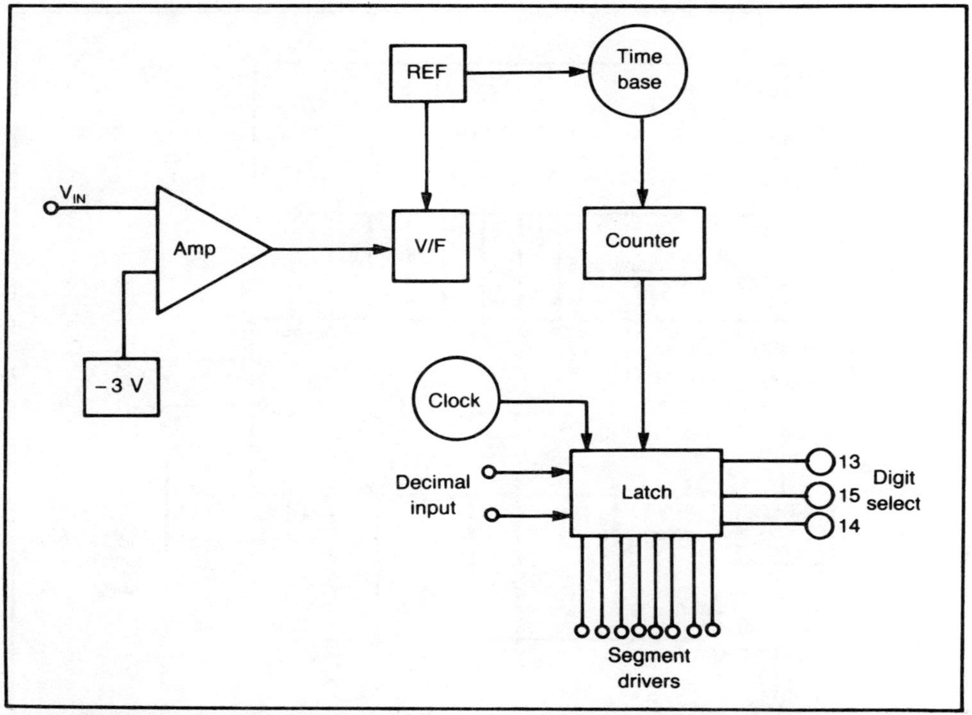

Fig. 12-37. The DM7700 contains almost all of the circuitry for a digital voltmeter, except for the display itself.

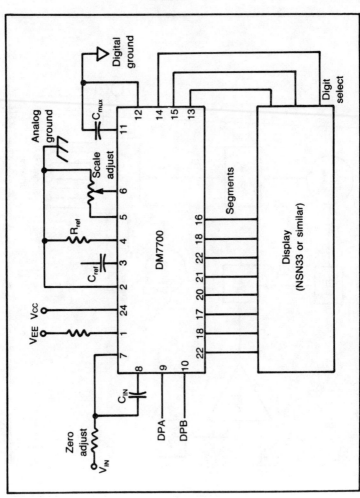

Fig. 12-38. This is a typical digital voltmeter circuit utilizing the DM7700.

274

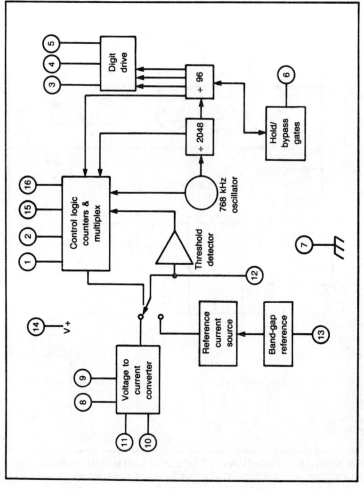

Fig. 12-39. The CA3162E is an A/D converter for use in digital voltmeters.

ure in traditional test instruments. Analog circuitry is really only suited for average ac measurements. Unfortunately, the average measurement often isn't very meaningful. The technician is usually more concerned with the rms (root mean square) values. An rms ac value corresponds to the same dc value. Rms values can be used with Ohm's law, while average values cannot.

In addition, the relationship between the average and the rms value varies according to waveshape. For simple sine waves, the rms value is about 1.11 times average value. This correction factor is used on analog meters, for direct read-out of the rms value if, and only if, the input is a pure sine wave. If the input is of any other waveshape, the reading is pretty much meaningless.

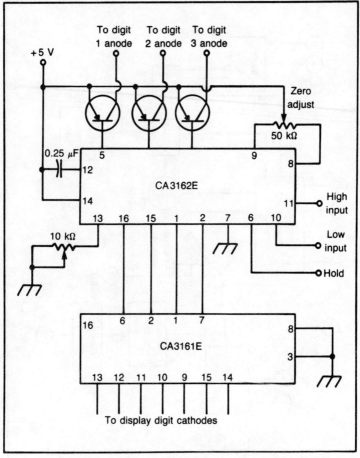

Fig. 12-40. The CA3126E/CA3161E set can be used to build a complete digital voltmeter with the addition of a display and a few external components.

276

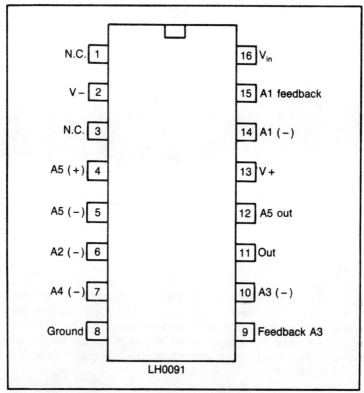

Fig. 12-41. The LH0091 is a true rms-to-dc converter circuit in IC form.

National Semiconductor's LH0091 is a true rms to dc converter. The output is an equivalent dc voltage which may be easily measured. A pinout diagram of this device is shown in Fig. 12-41.

Chapter 13

Digital Devices

```
     ┌───────┐
   ┌─┤1  S  24├─┐
   ┌─┤2  P  23├─┐
   ┌─┤3  E  22├─┐
   ┌─┤4  C  21├─┐
   ┌─┤5  I  20├─┐
   ┌─┤6  A  19├─┐
   ┌─┤7  L  18├─┐
   ┌─┤8  P  17├─┐
   ┌─┤9  U  16├─┐
   ┌─┤10 R  15├─┐
   ┌─┤11 P  14├─┐
   ┌─┤12 O  13├─┐
        S
        E
```

S O FAR IN THIS BOOK WE HAVE BEEN CONCENTRATING ON SPECIAL-purpose ICs for linear applications. Many of the chips we have discussed in the previous chapters have incorporated digital circuitry, and many of them were designed to be interfaced with digital devices. But the end result for the chips described through Chapter 12 has been some kind of analog signal.

Starting with this chapter, we will deal with true digital ICs, which are intended primarily for use in purely digital applications. Several of the devices we will examine in the next few chapters can also be used in certain linear applications, but we have to draw a line somewhere. The distinction between linear and digital ICs is growing more and more arbitrary.

SCHMITT TRIGGERS

One important type of digital IC is the Schmitt trigger. Of course, it can be used for linear switching applications, or for driving digital gates with analog signals. But the prime function of the Schmitt trigger is to clean up noisy digital signals.

A Schmitt trigger is primarily an electronic switch. When the input voltage exceeds a specific level, the output will go HIGH. When the input drops below a certain point, the output will go LOW.

Most Schmitt triggers have some kind of hysteresis. This means that the LOW switching point is lower than the HIGH switching point. This prevents the output from oscillating between states when the input is near one of the switching values.

A properly designed Schmitt-trigger circuit can pick out the desired digital signals, while ignoring most noise pulses that might show up on the line. The result is a clean string of digital pulses at the output, as illustrated in Fig. 13-1. Of course, a Schmitt trigger may not be able to help with severe noise problems, as illustrated in Fig. 13-2, but it can take care of most common "garden variety" noise.

Many Schmitt trigger ICs are available. Such devices are manufactured in all of the major logic families. The 7413 is a TTL device. It contains two independent NAND type Schmitt triggers with four inputs each in a single 14-pin package. This chip is illustrated in Fig. 13-3. Another TTL Schmitt trigger IC is the 7414, which is shown in Fig. 13-4. This chip consists of six independent single-input Schmitt-trigger stages.

Figure 13-5 illustrates a fairly typical Schmitt-trigger application. The analog signal on the base of the transistor controls the Schmitt trigger. This circuit is a simple one bit A/D (analog-to-digital) converter.

A somewhat less obvious application is the circuit shown in Fig. 13-6. Here a feedback path is added to the Schmitt trigger, forcing it to act as an oscillator. The frequency and the duty cycle of the output signal are determined by the values of the resistor and the capacitor. A wide range of frequencies can be achieved with a single capacitance value, just by varying the resistance. For example, using a 0.1 μF capacitor gives a frequency range from about 70 Hz to almost 300 kHz (300,000 Hz), depending on the value of the resistor.

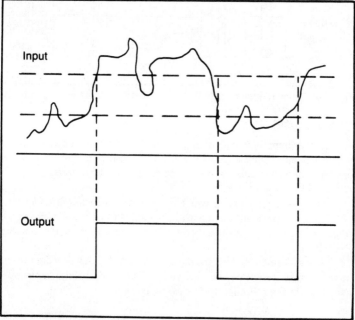

Fig. 13-1. A Schmitt trigger can clean up a noisy digital signal.

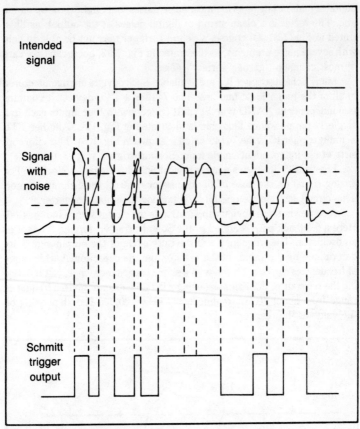

Fig. 13-2. Schmitt triggers can't always take care of severe noise problems.

SHIFT REGISTERS

Another important type of digital circuit is the shift register. This is sort of a memory device. It consists of several stages. Each stage can hold one bit. The individual bits are passed along from stage to stage, rather like a bucket brigade.

There are several ways in which shift registers are classified. One type of classification is defined by the way the data bits are passed through the stages. In the following examples we will start out with the binary number 101011.

Some shift registers shift the digits to the left. As each new digit opens up to the right, it is cleared (loaded with 0). That is, if we enter our sample number, it will go through the following pattern:

        101011   original number
        010110

280

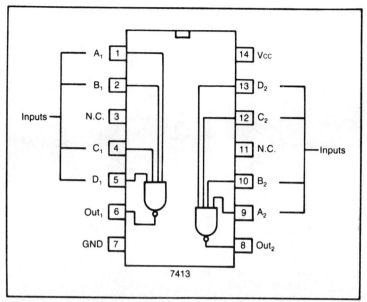

Fig. 13-3. The 7413 is a TTL with two independent NAND type Schmitt triggers with four inputs each.

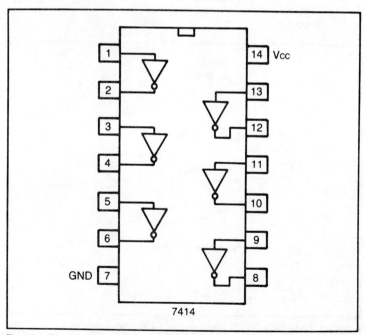

Fig. 13-4. The 7414 contains six independent single input Schmitt-trigger stages.

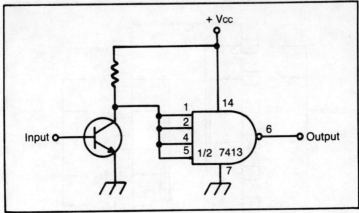

Fig. 13-5. A Schmitt trigger can be used as a simple 1-bit A/D converter.

```
101100
011000
110000
100000
000000
000000          register is now cleared—no further
000000          changes will take place
000000
```

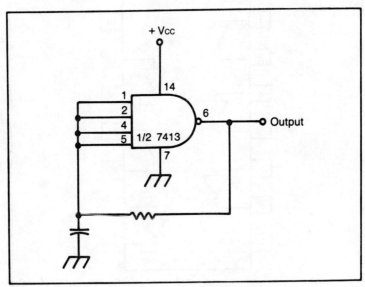

Fig. 13-6. Adding a feedback path can force a Schmitt trigger to act as an oscillator.

A variation on the shift-left type shift register loops the displaced digits from the left back around to fill the newly opened places on the right. Once again, starting out with the same sample number, we get the following pattern:

101011	original number
010111	
101110	
011101	
111010	
110101	
101011	the original number again
010111	
101110	
011101	
111010	

This pattern will repeat indefinitely.

Other shift registers move the bits in the opposite direction. That is, they shift right. Again, some fill the newly opened spaces to the left with zeroes:

101011	original number
010101	
001010	
000101	
000010	
000001	
000000	
000000	register is now cleared
000000	

Some right shift registers loop the displaced digits back around to the left.

101011	original number
110101	
111010	
011101	
101110	
010111	
101011	the original number again
110101	
111010	
011101	

And once more, the cycle will be endlessly repeated.

Some shift registers are bidirectional. That is, they can shift left or shift right.

Shift registers are also classified according to how the digits are presented at the inputs and outputs. Serial data is one bit at a time. Parallel data transmits the entire word (all bits) at once.

There are four possible combinations;

SISO	Serial In—Serial Out
SIPO	Serial In—Parallel Out
PISO	Parallel In—Serial Out
PIPO	Parallel In—Parallel Out

Some shift registers may operate in more than one of these modes. For example most PIPO shift registers could be used as SISO shift registers, merely by using just one bit at a time. Table 13-1 lists a few TTL and CMOS shift register ICs. Some typical devices are illustrated in Figs. 13-7 through 13-12.

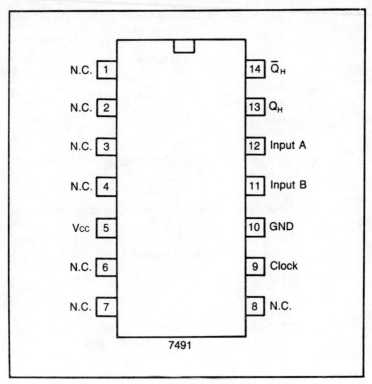

Fig. 13-7. The 7491 is an 8-bit TTL SISO shift-register IC.

Table 13-1. Here Is a Comparison of a Few Typical TTL and CMOS Shift Register ICs.

Type	Function	# of Bits	Shift Right	Shift Left	Bi?	Logic Family	Fig.	Freq. (MHz)	Load	Hold
7491	SISO	8	YES	NO	NO	TTL	13-7	10	NO	NO
7494	SISO	4	YES	NO	NO	TTL	—	10	YES	NO
7495	PIPO	4	YES	NO	NO	TTL	13-8	36	YES	NO
7496	PIPO	5	YES	NO	NO	TTL	—	10	YES	NO
74164	SIPO	8	YES	NO	NO	TTL	13-9	25	NO	YES
74165	PISO	8	YES	NO	NO	TTL	13-10	25	YES	YES
74166	PISO	8	YES	NO	NO	TTL	—	20	YES	YES
74178	PIPO	4	YES	NO	NO	TTL	—	25	YES	YES
74179	PIPO	4	YES	NO	NO	TTL	—	25	YES	YES
74194	PIPO	4	YES	YES	YES	TTL	—	25	YES	NO
74195	PIPO	4	YES	NO	NO	TTL	—	30	YES	YES
74198	PIPO	8	YES	YES	YES	TTL	—	25	YES	YES
74199	PIPO	8	YES	NO	NO	TTL	—	25	YES	YES
4006	SISO	18	YES	NO	NO	CMOS	13-11	10	NO	NO
4014	PISO	8	YES	NO	NO	CMOS	—	5	YES	NO
4015	SIPO	8	YES	NO	NO	CMOS	—	9	NO	NO
4021	PISO	8	YES	NO	NO	CMOS	—	5	YES	NO
4031	SISO	64	YES	NO	NO	CMOS	—	8	NO	NO
4034	PIPO	8	YES	YES	YES	CMOS	13-12	10	YES	YES
4035	PIPO	4	YES	NO	NO	CMOS	—	5	YES	NO
4094	SIPO	8	YES	NO	NO	CMOS	—	5	NO	YES
40100	SISO	32	YES	YES	YES	CMOS	—	3	NO	NO
40104	PIPO	4	YES	YES	YES	CMOS	—	9	YES	YES
40194	PIPO	4	YES	YES	YES	CMOS	—	9	YES	YES

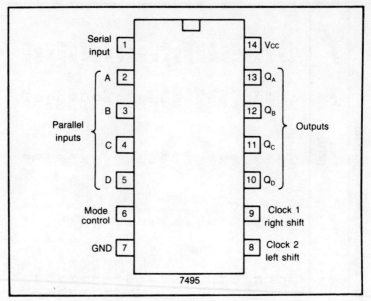

Fig. 13-8. Another TTL shift-register IC is the 4-bit PIPO 7495.

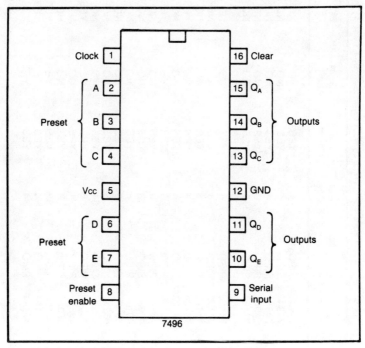

Fig. 13-9. A typical SIPO shift register is the 8-bit 7496.

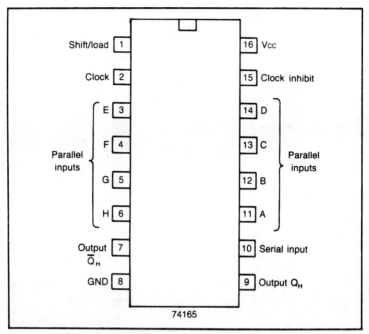

Fig. 13-10. The 74165 is an 8-bit PISO shift register.

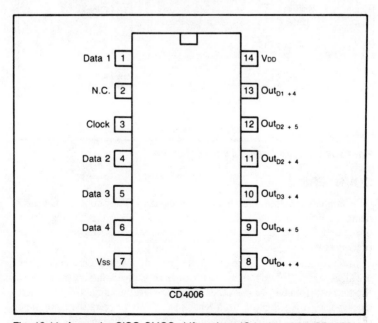

Fig. 13-11. A popular SISO CMOS shift register IC is the 18-bit CD4006.

287

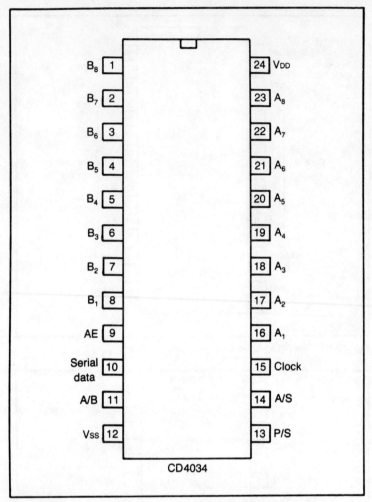

Fig. 13-12. The CD4034 is an 8-pin CMOS PIPO shift register IC.

COUNTERS

Counters are closely related to shift registers. They do just what their name implies—count. Simple counters are made up of flip-flop stages, as illustrated in Fig. 13-13. Each successive stage is triggered by its predecessor.

Initially all of the flip-flop stages are cleared, or reset. This means they are all at logic 0. When the first clock pulse is received, it triggers the first flip-flop (A) from 0 to 1. The other stages are left unchanged. On the next clock pulse, A is triggered from 1 back to 0. This triggers the next stage (B) from 0 to 1. The same procedure continues, following this pattern:

Clock Pulse #	A	B	C	D	Count	
0	0	0	0	0	0000	(0)
1	1	0	0	0	0001	(1)
2	0	1	0	0	0010	(2)
3	1	1	0	0	0011	(3)
4	0	0	1	0	0100	(4)
5	1	0	1	0	0101	(5)
6	0	1	1	0	0110	(6)
7	1	1	1	0	0111	(7)
8	0	0	0	1	1000	(8)
9	1	0	0	1	1001	(9)
10	0	1	0	1	1010	(10)
11	1	1	0	1	1011	(11)
12	0	0	1	1	1100	(12)
13	1	0	1	1	1101	(13)
14	0	1	1	1	1110	(14)
15	1	1	1	1	1111	(15)
16	0	0	0	0	0000	(0)
17	1	0	0	0	0001	(1)
18	0	1	0	0	0010	(2)
19	1	1	0	0	0011	(3)
20	0	0	1	0	0100	(4)

This pattern can continue indefinitely.

Notice that when the maximum count is exceeded (15 in this case) the counter clears itself back to zero, and starts over. The maximum count a counter circuit is capable of is referred to as the *modulo* of that counter. Counters with modulos equal to any whole number greater than one can be constructed by using a string of flip-flops as described above. The four-stage counter in our example has a modulo of 16. That is, there are 16 count steps from 0000 (0) to 1111 (15).

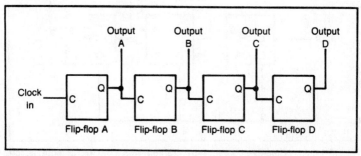

Fig. 13-13. Simple counters are made up of flip-flop stages.

For modulos that are powers of two (4, 8, 16, etc.) the design is quite simple. All that such a counter needs is a string of the appropriate number of flip-flop stages—one for each digit in the outputted binary number (or, for each power of two).

But what if, for example, we need a counter with a modulo of five? Five is definitely not a power of two. Most flip-flops have a clear (or C or R or reset) input that can force the Q output back to logic 0. By using some digital gates, we can force the flip-flop stages to clear after a specific count is reached. For a modulo five counter, we want the output count sequence to run like this:

$$000$$
$$001$$
$$010$$
$$011$$
$$100$$
$$000$$
$$001$$

and so on.

The output count must be forced back to 000 after a count of 100. The counter needs three binary digits for all of its valid count values. The first step in setting up a modulo-five counter is to start with a modulo-eight counter, since eight is the next highest power of two. This requires three flip-flop stages, as illustrated in Fig. 13-14. Now, we want the counter to reset after the output count passes:

$$A = 0$$
$$B = 0$$
$$C = 1$$

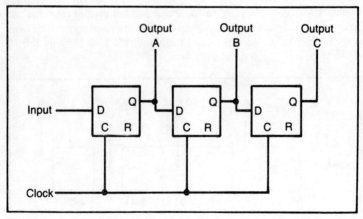

Fig. 13-14. A modulo-five counter must start out with a modulo of eight, which is the next higher power of two.

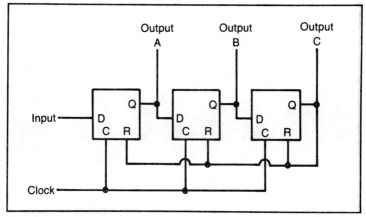

Fig. 13-15. By forcing the reset before the maximum count is reached, any integer modulo less than the maximum can be set up.

This happens to be the only valid output combination in which C is at logic 1. Therefore, we can simply feed this output line back to the clear inputs of the flip-flops, as shown in Fig. 13-15. Even though the flip-flops are theoretically capable of counting to 111 (decimal 7), the feedback prevents the counter from ever exceeding 100 (decimal 4). The result is a modulo-5 counter.

Now let's change this to a modulo-6 counter. This is only slightly more complex than the previous problem. In this case, we want the outputs to cycle through this sequence:

$$
\begin{array}{c}
000 \\
001 \\
010 \\
011 \\
100 \\
101 \\
000 \\
001
\end{array}
$$

and so on.

This time the counter resets after a count of 101. For this problem, we can't just feed back the C output to the clear inputs, or the counter will never get a chance to put out the desired count step of 101. We need to reset the flip-flops when both output A and output C are at logic 1. (Note that output B doesn't matter, since the count of 111 will never be reached). Clearly we can solve this problem with a simple AND gate. Outputs A and C are ANDed together, and the result is used to drive the clear inputs. This circuit is illustrated in Fig. 13-16.

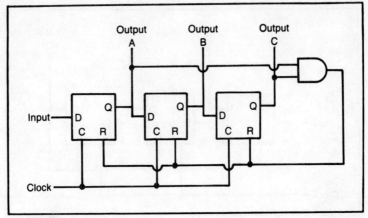

Fig. 13-16. For some modulos, external gating may be required to define the forced reset point.

By using the right combination of gates literally any integer modulo can be achieved. Some are slightly more complex than the simple examples shown here—especially when relatively large modulo values are involved. But the same simple principle applies.

In most practical applications you won't want to bother with individual flip-flop stages. Many advanced counters are available in convenient IC form. Generally, they can be used in much the same way as the flip-flop counters already described.

The 74161 is a high speed 4-bit binary counter IC. This TTL chip is illustrated in Fig. 13-17. It is a simple 4-stage counter, much like the circuits we have just looked at. It is designed to count very high speed clock frequencies. Input frequencies up to 25 MHz (25,000,000 Hz) can be handled by the 74161.

Another TTL counter IC is the 7490, which is shown in Fig. 13-18. This device is a 4-bit ripple type decade counter. A decade counter is designed for a modulo of ten. Since ten is the base of the number system we normally work with, decade counters are extremely handy for applications in which the count is to be displayed numerically.

The 7490 is divided into two counter sections. One is a divide-by-two counter, and the other is a divide-by-five counter. Of course, if the two are used together, the result will be a divide-by-ten counter. But other combinations are also possible, making the 7490 a very versatile device.

For BCD (binary-coded decimal) counting, the incoming clock pulses to be counted are fed to input A (pin 14). Output Q_0 (pin 12) is fed into input B (pin 1). The BCD outputs are taken off Q_0 (pin 12), Q_1 (pin 9), Q_2 (pin 8), and Q_3 (pin 11).

In a symmetrical Bi-quinary divide-by-ten counter (frequency divider), the input signal is fed to input B (pin 1). The Q_3 output (pin 11) is fed back to input A (pin 14). A square wave with one tenth the frequency as the input signal will then be available at the Q0 output (pin 12).

The divide-by-two and divide-by-five counters may also be used independently. In this case, no external interconnections are required. The divide-by-two counter uses input A (pin 14), and output Q0 (pin 12), while the divide by five counter employs input B (pin 1), and outputs Q1 (pin 9), Q2 (pin 8), and Q3 (pin 11).

A basic divide-by-ten counter using the 7490 is illustrated in Fig. 13-19. Notice that pins 2, 3, 6, 7, and 10 are all grounded. For normal decade operation either pin 2 or pin 3 and either pin 6 or pin 7 must be grounded. It's easiest just to ground all of them. We will discover the significance of these pins shortly. The basic truth tables for the 7490 are given in Table 13-2.

The 7490 is triggered by a negative-going input pulse. A HIGH output appears on pin 12 on the first count. On the second count, pin 12 goes LOW, and pin 9 goes HIGH. On the third count both pin 12 and pin 9 go HIGH. The four output pins represent a BCD count.

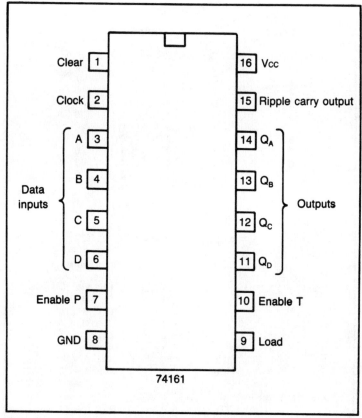

Fig. 13-17. The 74161 is a high speed 4-bit binary counter IC.

Table 13-2. These Are the Basic Truth Tables for the 7490.

BCD Count Sequence

Count	Outputs			
	QD	QC	QB	QA
0	0	0	0	0
1	0	0	0	1
2	0	0	1	0
3	0	0	1	1
4	0	1	0	0
5	0	1	0	1
6	0	1	1	0
7	0	1	1	1
8	1	0	0	0
9	1	0	0	1

Bi-Quinary

Count	Outputs			
	QD	QC	QB	QA
0	0	0	0	0
1	0	0	0	1
2	0	0	1	0
3	0	0	1	1
4	0	1	0	0
5	1	0	0	0
6	1	0	0	1
7	1	0	1	0
8	1	0	1	1
9	1	1	0	0

Reset/Count Truth Table

Reset Inputs				Outputs			
R0(1)	R0(2)	R9(1)	R9(2)	Qd	Qc	Qb	Qa
1	1	0	x	0	0	0	0
1	1	x	0	0	0	0	0
x	x	1	1	1	0	0	1
x	0	x	0	Count			
0	x	0	x	Count			
0	x	x	0	Count			
x	0	0	x	Count			

x = don't care

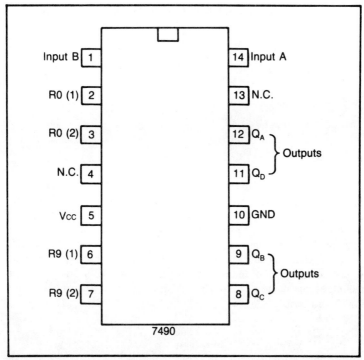

Fig. 13-18. The 7490 decade counter is useful in many applications in which the count is to be displayed.

Pin 2 and 3 are reset pins. When both are made HIGH, the counter resets to 0000. For normal counting, at least one of these reset pins must be LOW. These pins allow the circuit designer to change the modulo of the 7490.

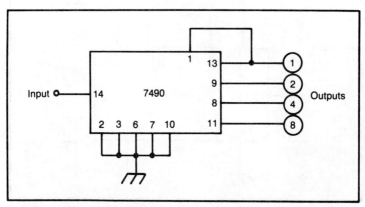

Fig. 13-19. This is a basic divide-by-ten counter circuit using the 7490.

Let's say we want a modulo-three counter. The desired count pattern is:

0000
0001
0010
0000
0001
0010
0000

and so on.

To get this pattern, the ground connections to pins 2 and 3 are removed. Pin 3 is connected to a permanent HIGH, or left floating (a floating TTL input acts like a HIGH). The Q1 (2) output (pin 9) is shorted to the other reset terminal (pin 2). When Q0 goes HIGH, both pins 2 and 3 are made HIGH, so the counter is reset to 0000. This circuit is shown in Fig. 13-20.

If we want a modulo that requires a reset after a count with two HIGH values, we use both of the reset pins. For instance, let's say we want a modulo-seven counter. This would require a reset after a count of six, giving this pattern:

0000
0001
0010
0011
0100

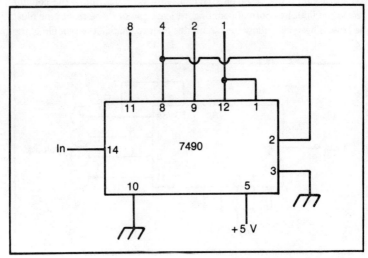

Fig. 13-20. The 7490 can use forced resetting for modulos less than ten.

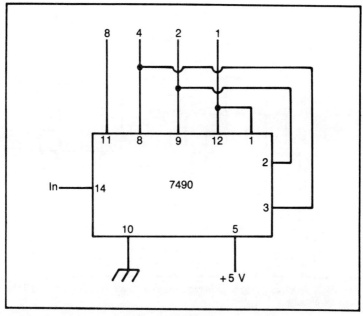

Fig. 13-21. The 7490 has two reset inputs, reducing the need for external AND gates.

0110
0000
0001
0010

and so on.

Both outputs Q1 (pin 9) and Q2 (pin 8) must go HIGH. Just connect pin 9 to reset pin 2, and pin 8 to reset pin 3, as shown in Fig. 13-21. Now the counter will be forced to reset to 0000 after a count of 0110.

For some modulos an external logic gate may be required. For example, a modulo-eight counter would require a reset after the count of 0111. Three of the outputs must go HIGH before the counter is reset. There aren't enough reset pins, so an external AND gate must be used to combine two of the outputs into a single logic signal.

The 7490 can be cascaded for multiple digit operation. When one 7490 passes its maximum count and resets, it applies a pulse to the next 7490 counter. This means a count of 09 is followed by 10, a count of 19 is followed by 20, and so forth.

Obtaining nonstandard modulos with multidigit 7490 counter circuits often requires multiple external gates, but the basic principle remains the same. A modulo-36 counter is illustrated in Fig. 13-22 as an example.

The 7492, which is shown in Fig. 13-23 is quite similar to the 7490,

297

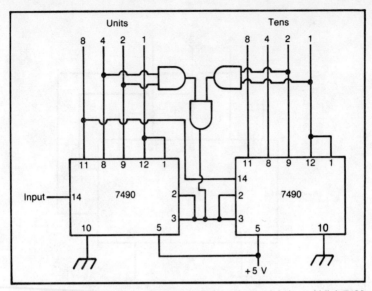

Fig. 13-22. External gating may be required to set the modulo in multidigit 7490 counter circuits.

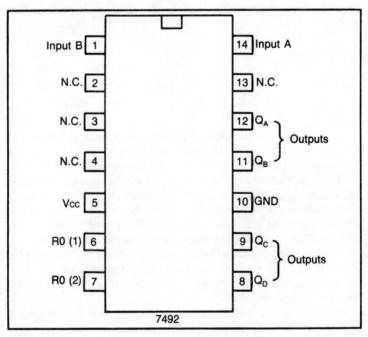

Fig. 13-23. The 7492 is similar to the 7490, except it is a divide-by-twelve counter.

298

except this chip is a 4-bit ripple type divide-by-12 counter. Like the 7490, the 7492 is made up of two independent counter stages—in this case a divide-by-two counter and a divide-by-six counter.

Another powerful TTL counter IC is the 74169, which is shown in Fig. 13-24. This is a 4-bit, modulo-sixteen counter. It can either count up or count down. The 74169 features synchronous operation. This means that all of the internal flip-flop stages change states simultaneously when triggered by the count enable inputs and internal logic gates. This prevents the noise and output spikes which often occur with asynchronous (ripple type) counters, like the 7490.

Another interesting feature of the 74169 is the fact that it is programmable. Each of the outputs may be preset to either a HIGH or LOW state. This means the count can start from any value within the counter's range.

Even more powerful counter ICs can be constructed using CMOS technology. Figure 13-25 shows the MC14553B. This is a three-decade BCD CMOS counter. The MC14553B contains three separate divide-by-ten counters. All three decade counters use the same four output pins (5, 6,

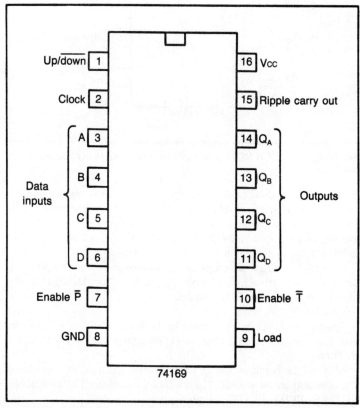

Fig. 13-24. The 74169 can count in either direction (up or down).

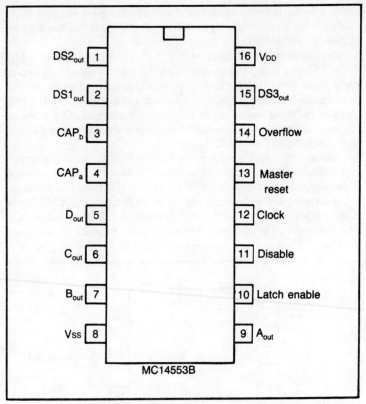

Fig. 13-25. The CMOS MC14553B contains three decade BCD counters in a single package.

7, and 9). A set of latches define which of the three decade counters is using the outputs. The displayed count is periodically updated and held on the associated digital readout, while the counter continues counting. Because of the somewhat unusual nature of this IC, I believe it would be helpful to take a pin-by-pin look at the MC14553B.

Pin 1—This output pin activates digit two (second decade output). The other two digits are selected by pin 2 (digit 1) and pin 15 (digit 3). The digit select outputs are TTL compatible.

Pin 2—Digit 2—see pin 1.

Pins 3 and 4—An external capacitor is connected between these two pins. This capacitor sets the frequency of the internal digit select multiplex oscillator.

Pin 5—This is output D (the most-significant bit) of the BCD counters. The other outputs are at pins 6, 7, and 9. These outputs are TTL compatible.

Pin 6—BCD output C—see pin 5.

Pin 7—BCD output B—see pin 5.

Pin 8—This pin is used for the V_{SS} power supply connection.

Pin 9—BCD output A (least-significant bit)—see pin 5.

Pin 10—This is the latch enable input. When this pin is made HIGH, the latch is loaded with the current output count value.

Pin 11—This is a disable input. Counting can only occur when this pin is held LOW.

Pin 12—This is the CLOCK input. The signal to be counted is fed into the counter at this terminal.

Pin 13—This pin is labelled MR, or *master reset*. For ordinary operation this pin is held LOW. If it is brought HIGH, all four BCD outputs will be brought LOW (logic 0). To preserve the last count in the latch, make sure LE (pin 10) is kept HIGH during reset operations.

Pin 14—The V_{DD} power supply connection is made to this pin.

Like most other counter ICs, two or more MC14553Bs can be cascaded to increase the count range. Since each MC14553B contains three decade counters, a cascaded pair would cover a six-decade range, for a modulo of 1,000,000.

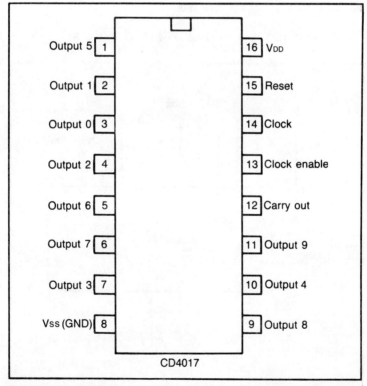

Fig. 13-26. The CD4017 counts out in a somewhat different way than the other counters described so far.

Some IC counters count out in a somewhat different way. An example of this is presented by another popular CMOS counter chip, the CD4017. The pinout of this device is shown in Fig. 13-26. This IC has ten outputs numbered from 0 to 9. On any specific count, only one of these outputs is HIGH, and the other nine are LOW. In other words, this is a one-of-ten counter.

By grounding pin 15 of the CD4017, and connecting pin 13 to one of the outputs, the counter will count from 0 to that output's value, and then stop. Figure 13-27 illustrates how this chip can be wired to count to seven and then stop. The circuit may be reset for another count cycle by temporarily disconnecting pin 15 from ground and momentarily connecting it to a positive voltage source. Generally, the VDD supply voltage will be used.

A slightly different application for the 4017 is illustrated in Fig. 13-28. Here the connections to pins 13 and 15 are reversed from Fig. 13-26. This time pin 13 is grounded, and pin 15 is connected to the maximum count output line. If the 4017 is wired as shown here, the outputs will count from 0 to 7, then revert back to 0 and start over. In other words, the outputs will sequentially go HIGH in this order:

0 1 2 3 4 5 6 7 0 1 2 3 4 5 6 7 01 2 3 . . .

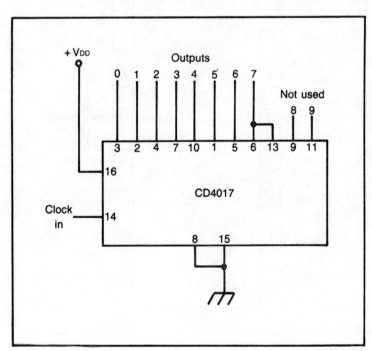

Fig. 13-27. This is a modulo-seven counter built around the CD4017.

302

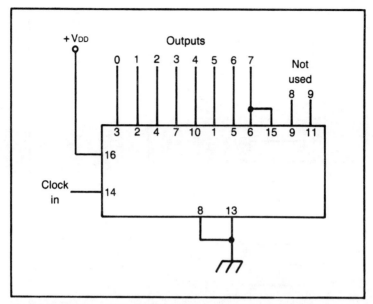

Fig. 13-28. By connecting the CD4017 in this way, it will continuously cycle through its count sequence.

Of course, any count up to 9 may be set up in this manner. But, now let's suppose we need a circuit that will count higher than nine? As you probably expect, the solution is to cascade two or more CD4017s, as shown in Fig. 13-29. This two 4017 circuit can count from 00 to 99. The first counter represents the one's column, and the second counter represents the ten's column.

If one of the feedback methods described above are used to create a modulo less than 100, a two-input gate will be needed, because two outputs (a one's output and a ten's output) are needed to uniquely indicate each number. Figure 13-30 shows a circuit that counts to 54, cycles back to 00 and repeats. This approach can easily be expanded for still higher counts. A third CD4017 will permit counts up to 999. A four CD4017 circuit has a maximum possible count of 9999.

An even more versatile CMOS counter IC is the CD4018, which is shown in Fig. 13-31. This chip is called a programmable counter. It can be operated in two modes. In one mode, the 4018 functions much like the 4017. It can divide (count) the input signal by any number from 2 to 10. A feedback loop is required for this circuit to operate, and the output is taken off pin 1 (DATA INPUT) which is in the feedback path. The output is a square wave (or near square wave—odd counts throw off the symmetry somewhat) that has a frequency equal to:

$$F_o = F_i/C$$

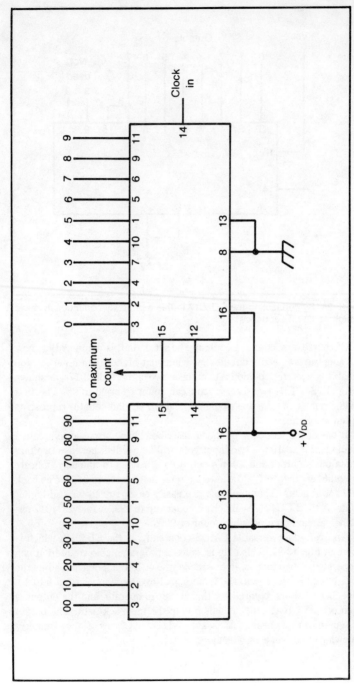

Fig. 13-29. Multiple CD4017s can be cascaded for longer count ranges.

304

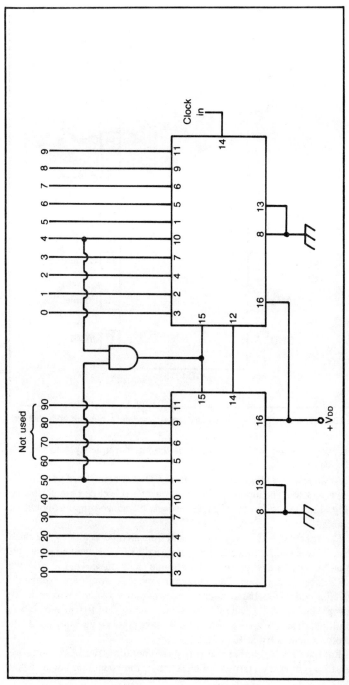

Fig. 13-30. This circuit will count to 54, then cycle back to 00 and repeat.

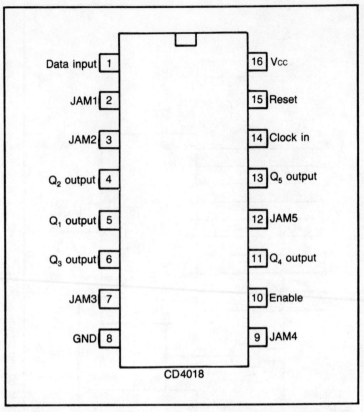

Fig. 13-31. The CMOS CD4018 is an extremely powerful programmable counter.

where F_o is the output frequency, F_i is the input frequency, and C is the count value.

As an example, Fig. 13-32 shows the CD4018 being used as a divide-by-five counter. The input and output signals for this circuit are illustrated in Fig. 13-3. As with most CMOS devices, all unused inputs should be grounded.

Each of the Q outputs is phase shifted from its predecessor by one clock pulse. This is similar to what happens with the 4017. However, each output on the 4017 stays HIGH for only a single clock pulse. On the 4018 we have a different situation. The output duty cycle is always approximately 50%. For odd counts the symmetry of the output waveform will be off by one input clock pulse. Dividing by any number from 2 to 10 (or even higher, if multiple CD4018s are cascaded) will always produce virtually the same waveform symmetry at the output.

For the CD4018's other major mode of operation, the JAM inputs are used to program the starting count. That is, the count can begin at any

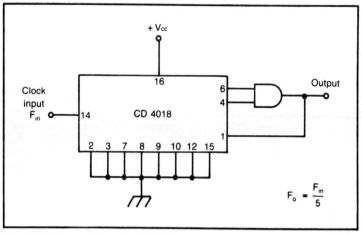

Fig. 13-32. Here we have the CD4018 being used as a divide-by-five counter.

value between zero and the maximum count value. For example, you could set up a counting sequence like this:

$$3 \quad 4 \quad 5 \quad 6 \quad 7 \quad 8 \quad 3 \quad 4 \quad 5 \quad 6 \quad 7 \quad 8 \quad 3 \quad 4 \quad 5 \quad \ldots$$

Jam inputs are loaded in parallel fashion, much like loading a parallel-input shift register.

Ordinarily, digital circuits work with rectangle waves only. Interestingly enough, the CD4018 counter can actually generate sine waves, and other analog waveshapes.

The CD4018 provides several phase-shifted outputs. Each is delayed by exactly one input (clock) pulse. If we sum the outputs together with the correct relative weighting, the result will be a staircase wave (like the

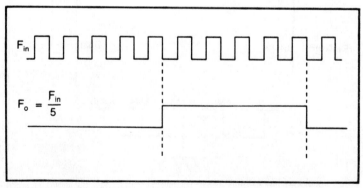

Fig. 13-33. These are the input and output signals for the circuit shown in Fig. 13-32.

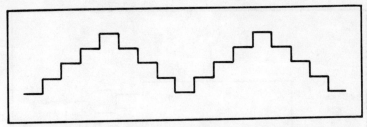

Fig. 13-34. If we sum the CD4018's outputs together, we get a staircase wave.

one illustrated in Fig. 13-34). External low-pass filtering can smooth out the steps to create an analog waveform. The basic circuit is shown in Fig. 13-35.

The output waveshape is defined by the relative weighting of each output. The weighting in turn, is determined by the resistor values. If the resistors have equal values, the output will more or less resemble a triangle wave.

A sine wave pretty much flattens out at its peaks. A triangle wave (and the digital waveform of Fig. 13-34) have fairly sharp peaks. We can modify the circuit, as shown in Fig. 13-36 to get flatter peaks. We have eliminated output Q5 from the output, even though it is still part of the counting cy-

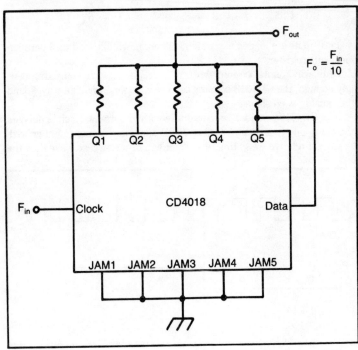

$$F_o = \frac{F_{in}}{10}$$

Fig. 13-35. This circuit can be used to generate a rough digital approximation of a sine wave.

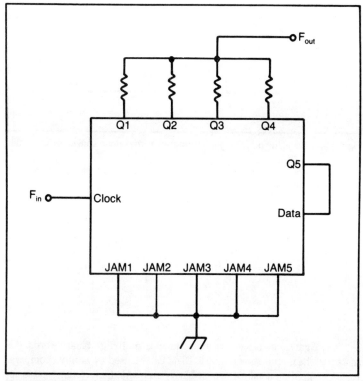

Fig. 13-36. This circuit will generate flatter peaks than the one shown in Fig. 13-35.

cle. As a result, the Q4 peak is effectively held longer, as shown in Fig. 13-37. Filtering this signal gives us a reasonably smooth pseudo-sine wave, as illustrated in Fig. 13-38. This isn't an ideal solution, however, since one less output is included in the waveform, the resolution has been reduced. That is instead of five steps to the peak, there are only four. Figure 13-39 shows a practical CD4018 digital sine-wave generator circuit.

THE DIGITAL COMPARATOR

Back in Chapter 12, we looked at analog comparators. These devices look at two signals (input and reference), and determine which is the larger. The concept of a digital comparator should be fairly obvious. Digital circuits work with numbers, so a digital comparator compares two binary numbers and determines which is the larger value.

A simple digital comparator is simply an X-OR (exclusive OR) gate. See Fig. 13-40. This type of gate gives a HIGH output when one and only one of the two inputs is HIGH. If both inputs are LOW, or if both are HIGH, the output will be LOW. In other words, the output is LOW (logic 0) when

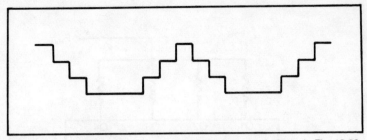

Fig. 13-37. This is the output signal generated by the circuit shown in Fig. 13-36.

the inputs are equal, and HIGH (logic 1) if they are unequal. In truth table form, the X-OR gate looks like this:

Inputs	Output
A B	
0 0	0
0 1	1
1 0	1
1 1	0

X-OR gates can be combined to compare multi-digit binary words. For instance, the circuit shown in Fig. 13-41 can be used as a digital comparator for comparing a pair of two-bit binary numbers. Unfortunately, this approach quickly gets awkward as the number of bits increases. Moreover, the X-OR gate doesn't give any indication of which of the inputs is larger. It merely indicates equality or nonequality.

Where there's a need, there's a special purpose IC. Dedicated digital comparator ICs are available. They are usually called magnitude comparators. Figure 13-42 shows a typical device of this type. It is the TTL 7485 4-bit magnitude comparator. This chip has three fully decoded outputs, which can indicate which of the 4-bit input values is the larger, or if the two are equal. Multiple 7485s can be cascaded to compare binary numbers of eight bits or larger.

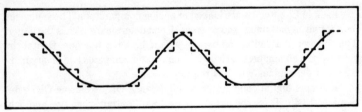

Fig. 13-38. Filtering the output of the circuit shown in Fig. 13-36 produces a fair approximation of a sine wave.

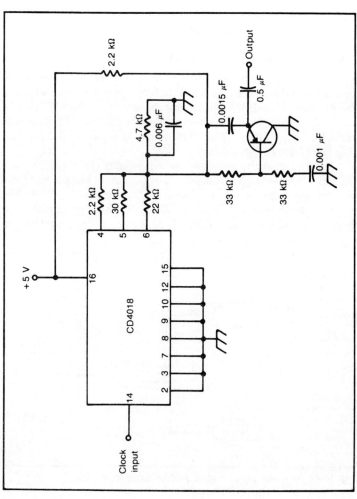

Fig. 13-39. This is a practical digital sine-wave generator circuit, built around the CD4018.

311

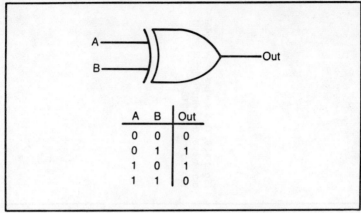

Fig. 13-40. A simple 1-bit digital comparator is an X-OR gate.

PROGRAMMABLE LOGIC GATES

The basic unit of digital circuits is the gate. Most gates have two inputs and one output. There are several basic types. Figure 13-43 shows the AND gate. The NAND gate is just the opposite of the AND gate. It is shown in Fig. 13-44. In Fig. 13-45 we have the OR gate, and its opposite is the NOR gate, shown in Fig. 13-46. The X-OR (exclusive OR) gate was illustrated back in Fig. 13-40.

Any truth table can be created by combining the basic gates in various ways. In many cases this may be awkward, but the demand isn't sufficient for any manufacturer to put out a dedicated special logic chip. The solution is a sort of do-it-yourself logic chip. Two basic types are the *field-programmable-logic array* (FPLA) and the programmable-logic array (PAL—a trademark of Monolithic Memories, Inc.).

These devices are similar to programmable-read-only memories

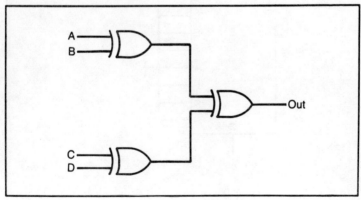

Fig. 13-41. This is a digital comparator for two 2-bit words.

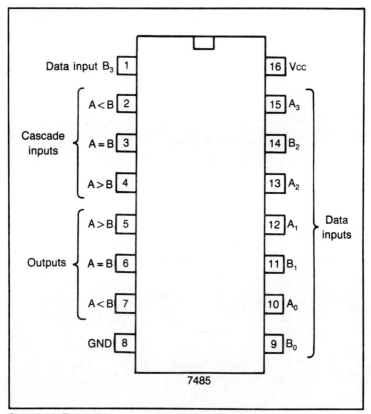

Fig. 13-42. The 7485 is a 4-bit magnitude comparator.

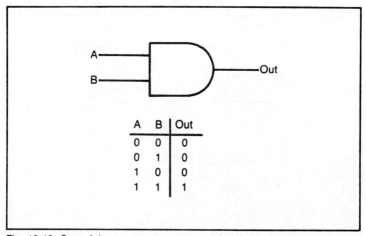

Fig. 13-43. One of the most common types of digital gate is the AND gate.

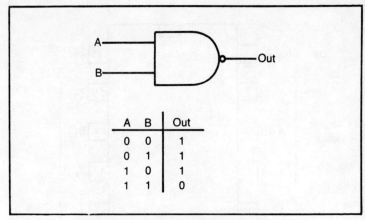

Fig. 13-44. A NAND gate is just the opposite of an AND gate.

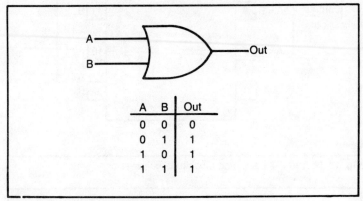

Fig. 13-45. Another common digital gate is the OR gate.

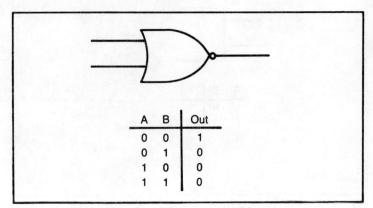

Fig. 13-46. The opposite of an OR gate is the NOR gate.

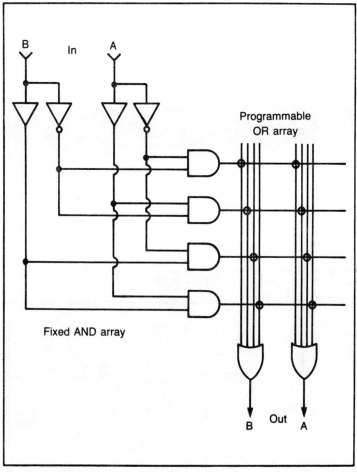

Fig. 13-47. A PROM contains fusible links which can be permanently programmed by the user.

(PROM). The user can permanently burn in any data he chooses, by applying a voltage to internal fusible links. Once programmed, the data cannot be changed.

A simplified block diagram of a typical PROM is shown in Fig. 13-47. The solid dots indicate permanent connections made by the manufacturer. The small circles indicate the fusible links, which are user programmable. Notice that the AND inputs are fixed, and the OR inputs are programmable.

A PAL works in pretty much the opposite way as a PROM. As Fig. 13-48 illustrates the AND inputs are user programmable and the OR inputs are fixed in the PAL.

The FPLA is the most versatile of this type of device. As shown in

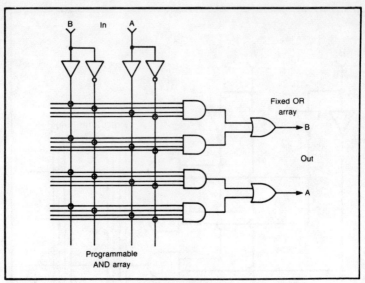

Fig. 13-48. A PAL is quite similar to a PROM.

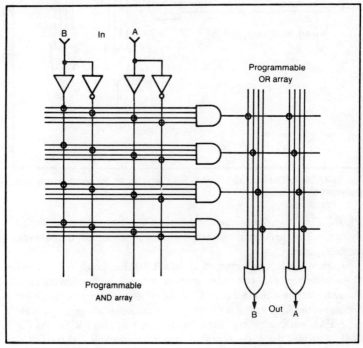

Fig. 13-49. A FPLA is more versatile than either the PROM (Fig. 13-47), or the PAL (Fig. 13-48).

316

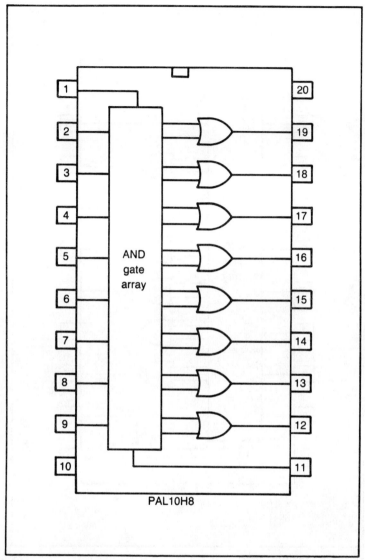

Fig. 13-50. The PAL10H8.

Fig. 13-49, both the AND inputs and the OR inputs are programmable in the FPLA. Unfortunately, a FPLA is usually more expensive and more difficult to program than either a PROM or a PAL. Some typical PAL chips are illustrated in Fig. 13-50 (the PAL10H8), Fig. 13-51 (the PAL12H6), and Fig. 13-52 (the PAL16L8). A number of other similar devices are also available.

317

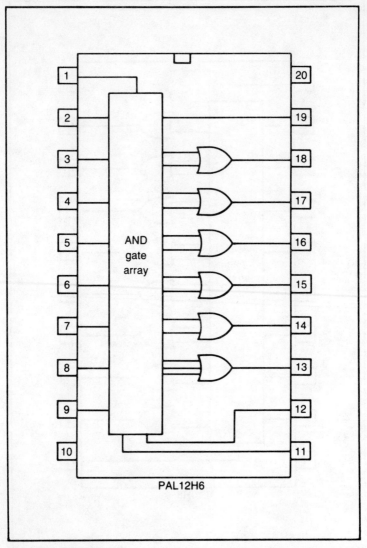

Fig. 13-51. The PAL12H6.

Some PALs and FPLAs contain internal flip-flop latches to allow functions such as counting, shifting, and sequencing. Even PALs without internal flip-flop stages can duplicate almost any task that can be performed with a SSI *(small-scale integration)* or MSI *(medium-scale integration)* digital IC. A PAL can even duplicate a 4-bit ALU *(arithmetic logic unit)*. In some cases up to ten standard ICs can be replaced by a single PAL.

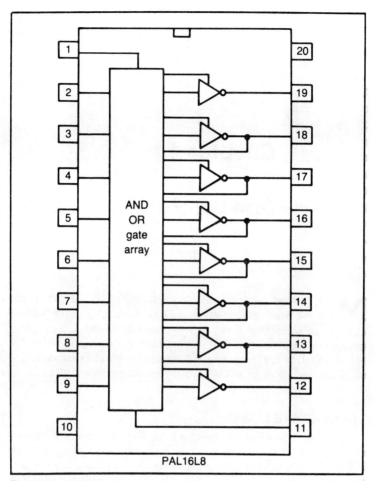

Fig. 13-52. The PAL16L8.

Chapter 14

Special-Purpose Logic

MOST DIGITAL DEVICES OPERATE ON SIMPLE BINARY LOGIC. THERE are two possible states—HIGH and LOW, 1 and 0, ON and OFF—these all mean the same thing. Binary logic is amazingly powerful, considering its fundamental simplicity. But there are some applications in which it is not sufficient. In this chapter we will look at a couple of variations on basic binary logic that are utilized in some special-purpose digital ICs.

THREE-STATE LOGIC

Ordinary binary logic is two-state logic. As long as power is applied to the circuit, all outputs are either HIGH or LOW. Some special-purpose ICs have a third possible output state. This is called, naturally enough, three-state logic. The third output state is a high impedance condition in which the output is effectively removed from the circuit. Why would we want to do this?

Consider the partial circuit shown in Fig. 14-1. Notice that several outputs are tied together into a single input. As long as all of these outputs are in the same state (either all HIGH, or all LOW) this circuit will work just fine. But if outputs A and B are HIGH, and outputs C and D are LOW, what is the logic state seen by the input? The input gets "confused," and circuit operation will be confused at best.

This type of arrangement is found in many computer systems, in which a number of devices (circuits) are connected to a single bus. Either all of the outputs have to always be at the same state (a very limiting requirement), or there must be some way to isolate all but one of the outputs from the bus. The solution is three-state logic. Three-state gates have an extra

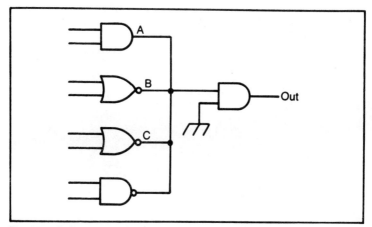

Fig. 14-1. Ordinary digital logic devices can run into problems when several outputs are connected to a single input.

enable input that switches the output into the high impedance (disconnected) state.

Figure 14-2 shows a basic three-state buffer. Essentially the enable input acts like a logic controllable switch, turning the output on and off. If the enable input is HIGH, the output will be in the same logic state as the input. If the enable input is LOW, the input will be disconnected from the output. The output will be in the high impedance state. The truth table for this device would look like this:

Inputs		Output
A	e	
x	0	Z
0	1	0
1	0	1

where A is the data input, and e is the enable input. An x represents "don't

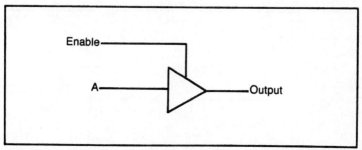

Fig. 14-2. The three-state buffer has a third high-impedance logic state.

care." The state of A is irrelevant when e = 0. Z stands for the high impedance output condition.

In some three-state gates, the enable input is inverted, as shown in Fig. 14-3. The truth table then looks like this:

Inputs		Output
A	e	
0	0	0
1	0	1
x	1	Z

For our discussion we have only looked at digital buffers, but any type of digital gate may use three-state logic.

MAJORITY LOGIC

Standard binary logic is rather rigid in its rules of operation. In some applications, we may prefer to have a more flexible response. For example, let's consider a security system with five sensors, but we only want the alarm to go off when at least three of the sensors are activated. It doesn't matter which three—just so three or more are triggered. If only one or two are activated, the alarm should not go off. We could conceivably accomplish this with standard logic gates, but as Fig. 14-4 shows, this can be rather awkward, at best.

Fortunately, there is an easier way. This easier way is called majority logic. Essentially, majority logic is like a democratic approach to gating. The majority of inputs rules. For a five input gate, at least three of the inputs must be in identical states. That is, at least three inputs will be HIGH, or at least three inputs will be LOW. There will always be a majority. A tie vote isn't possible with an odd number of inputs. The logic state of the majority of inputs determines the state of the output. This might sound a little confusing, but read on. It becomes clearer once you try a few examples.

The symbol for a majority gate is shown in Fig. 14-5. This particular

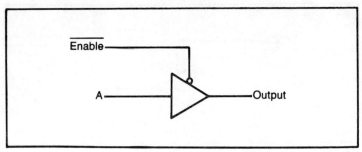

Fig. 14-3. In some three-state gates, the enable input requires an inverted signal.

322

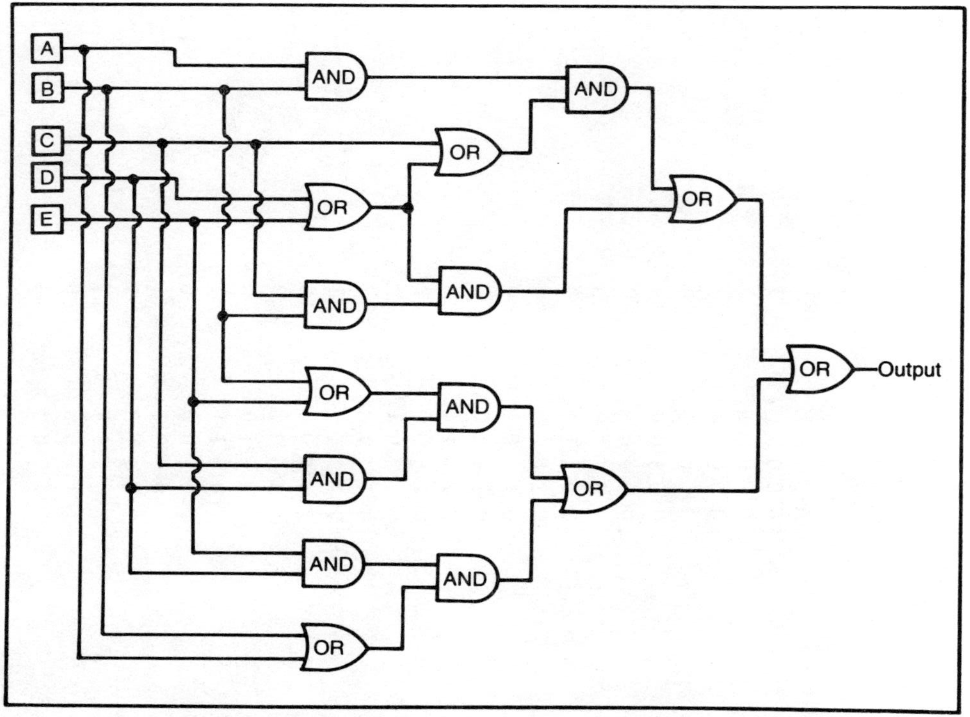

Fig. 14-4. Majority logic is inconvenient with ordinary digital gates.

323

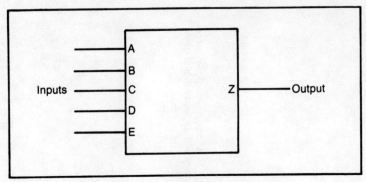

Fig. 14-5. This is the symbol for a five-input majority gate.

majority gate has five inputs (A, B, C, D, and E), and one output (Z). The truth table follows this pattern:

Inputs					Output
A	B	C	D	E	Z
0	0	0	0	0	0
0	0	0	0	1	0
0	0	0	1	0	0
0	0	0	1	1	0
0	0	1	0	0	0
0	0	1	0	1	0
0	0	1	1	0	0
0	0	1	1	1	1
0	1	0	0	0	0
0	1	0	0	1	0
0	1	0	1	0	0
0	1	0	1	1	1
0	1	1	0	0	0
0	1	1	0	1	1
0	1	1	1	0	1
0	1	1	1	1	1
1	0	0	0	0	0
1	0	0	0	1	0
1	0	0	1	0	0
1	0	0	1	1	1
1	0	1	0	0	0
1	0	1	0	1	1
1	0	1	1	0	1
1	0	1	1	1	1
1	1	0	0	0	0
1	1	0	0	1	1

324

Inputs	Output
1 1 0 1 0	1
1 1 0 1 1	1
1 1 1 0 0	1
1 1 1 0 1	1
1 1 1 1 0	1
1 1 1 1 1	1

Notice how for each input combination, the output (Z) goes along with the majority of the inputs.

A practical majority logic IC is the MC14530, which is shown in Fig. 14-6. This chip contains two five-input majority gates. It also has an X-OR gate connected to the output of each majority gate, and an extra input (W), as illustrated in Fig. 14-7. This makes the device even more versatile. If a logic 1 is applied to input W, the chip will function as a majority gate. If a logic 0 is fed to input W, the gate will function as a minority gate. In other words, the output of the majority gate will be inverted. This can be demonstrated by the following partial truth table for the X-OR stage.

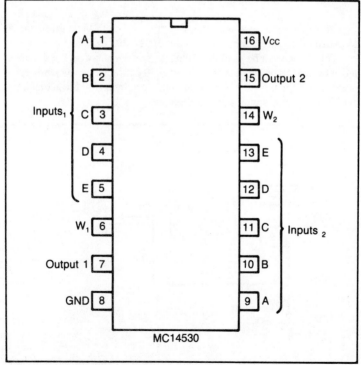

Fig. 14-6. The MC14530 contains two five-input majority gates.

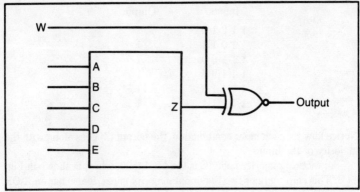

Fig. 14-7. The MC14530 also has X-OR gate stages for even more versatility.

W	Z	Output
0	0	1
0	1	0
1	0	0
1	1	1

Combine this truth table with the majority logic truth table given earlier to get the complete truth table for the MC14530.

This device can be used in a number of different ways. For example, it can be converted into a three-input majority gate, by permanently tying one of the main inputs HIGH, and a second LOW, then feed the actual input signals to the remaining three main inputs.

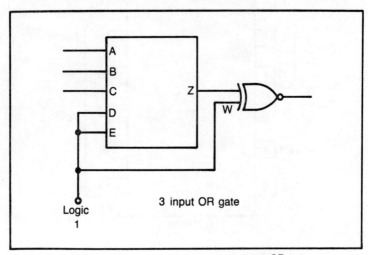

3 input OR gate

Logic
1

Fig. 14-8. The MC14530 can be wired as a three-input OR gate.

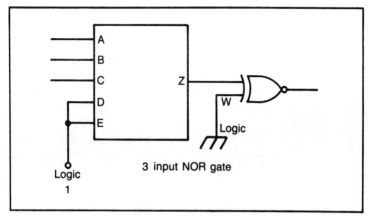

Fig. 14-9. A three-input NOR gate can also be simulated by the MC14530.

The MC14530 can easily be wired to duplicate most standard gating functions. For instance, Fig. 14-8 shows how this chip can be used as a three-input OR gate. Input W, and two of the main inputs are wired permanently HIGH. Inverting gates can be achieved simply by grounding input W. Figure 14-9 shows the MC14530 being used as a three-input NOR gate. Three input AND and NAND gates can be simulated by grounding (feeding a permanent logic LOW) to two of the main inputs.

Chapter 15

Multiplexers
and Demultiplexers

I N MANY COMPLEX DIGITAL SYSTEMS WE WILL NEED ONE SIGNAL AT
a given point in the circuit part of the time, but other signals will be
needed at the same point at other times. Conversely, a signal may need
to be routed through different paths at different times. These tasks can
be accomplished with circuits known as multiplexers and demultiplexers.

MULTIPLEXERS

To be able to select between different signal sources at a specific circuit point, we need a circuit called a multiplexer, or MUX, for short. Figure 15-1 illustrates how a simple multiplexer can be constructed from
standard NAND gates. This circuit has four main inputs (labeled 1 through
4) and two control inputs (labeled A and B). The logic signals fed to the
control inputs serve as address codes. They determine which of the main
input signals will reach the output. Only one of the main input lines is active at any one time. This can be seen in the truth table for this circuit:

Main Inputs				Control Inputs		Output
1	2	3	4	A	B	
0	x	x	x	0	0	0
1	x	x	x	0	0	1
x	0	x	x	0	1	0
x	1	x	x	0	1	1
x	x	0	x	1	0	0
x	x	1	x	1	0	1
x	x	x	0	1	1	0
x	x	x	1	1	1	1

328

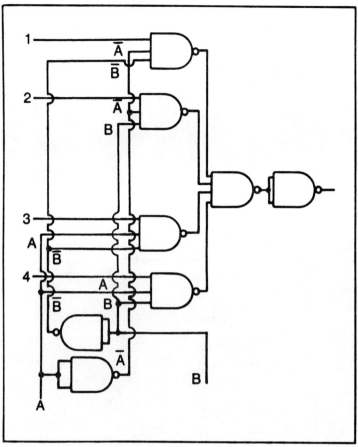

Fig. 15-1. Standard NAND gates can be combined to create a 1-of-4 multiplexer.

x represents "don't care." Any input marked x is simply ignored.

Notice that only one of the four main inputs lines has any influence on the output for any specific combination of values for control inputs A and B.

Because the control inputs can be used to select any of the main inputs, this type of circuit is sometimes called a data selector, although multiplexer is the preferred name.

The circuit shown in Fig. 15-1 is a 1-of-4 multiplexer, because any one of four inputs may be routed to the output. The same principle is commonly expanded to make 1-of-8 and 1-of-16 multiplexers. Multiplexers in all three of these standard sizes are readily available in IC form.

Some multiplexer ICs, it should be noted, also invert each of the main input signals (0s are changed to 1s, and vice versa) before feeding them to the output. In some applications, this may be an advantage, in others

it may turn out to be a disadvantage. Often it really won't make much difference one way or the other. This is the truth table for an inverting 1-of-4
multiplexer:

Main Inputs				Control Inputs		Output
1	2	3	4	A	B	
0	x	x	x	0	0	1
1	x	x	x	0	0	0
x	0	x	x	0	1	1
x	1	x	x	0	1	0
x	x	0	x	1	0	1
x	x	1	x	1	0	0
x	x	x	0	1	1	1
x	x	x	1	1	1	0

This is the same as the straight multiplexer truth table given above, of
course, except the output states are reversed for each input combination.

Multiplexers can be used in place of complex gating circuits. For example, consider this truth table:

Inputs				Output
A	B	C	D	
0	0	0	0	0
0	0	0	1	0
0	0	1	0	1
0	0	1	1	0
0	1	0	0	0
0	1	0	1	1
0	1	1	0	1
0	1	1	1	0
1	0	0	0	0
1	0	0	1	1
1	0	1	0	1
1	0	1	1	1
1	1	0	0	0
1	1	0	1	0
1	1	1	0	1
1	1	1	1	0

At best, it would be a definite nuisance to generate this truth table
using separate standard gates. The result would be a very dense circuit,
like the one shown in Fig. 15-2.

A 1-of-16 multiplexer can accomplish the same thing with a much simpler circuit. The 74150 is a typical 1-of-16 multiplexer in IC form. This
TTL device is illustrated in Fig. 15-3.

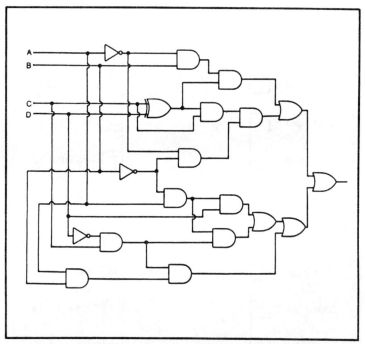

Fig. 15-2. Some truth tables require rather awkward and complex gating circuits.

Because there are 16 main, or data inputs, four control inputs (labeled A through D) are needed to uniquely identify each input address. These control inputs will correspond directly to the logic inputs in the truth table.

The 74150 happens to be an inverting type multiplexer, so we feed each main input with the opposite logic state desired for the output for the appropriate combination of control inputs. For example, when the control inputs are set to 1010, we want an output of logic 1, so we put the opposite state (logic 0) on input 10 (pin 21).

The complete circuit for generating the truth table in our example with a 74150 1-of-16 multiplexer IC is shown in Fig. 15-4. Notice how much simpler and easier this diagram is to follow than the comparable circuit in Fig. 15-2.

Literally any truth table can be readily generated using a multiplexer in this manner. Admittedly, in some cases it may actually be more convenient to use individual gates, but a multiplexer can come in very handy when complex and/or unusual truth tables must be generated. A 1-of-16 multiplexer like the 74150 can generate 2^{16} different truth tables. In other words, there are 65,536 possible combinations of inputs and outputs.

A multiplexer can also come in handy when unusual counting sequences are called for. For instance, the circuit shown in Fig. 15-5 combines a 4-bit modulo-16 counter with a 74150 1-of-16 multiplexer, producing the following

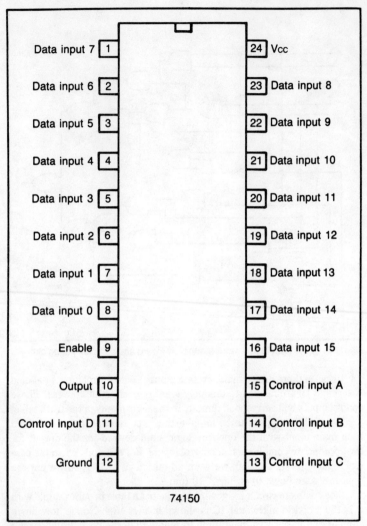

Fig. 15-3. The 74150 is a 1-of-16 multiplexer IC.

output pattern:

Clock Pulse	Counter Outputs	Output
1	0001	0
2	0010	1
3	0011	0
4	0100	0
5	0101	1

Clock Pulse	Counter Outputs	Output	
6	0110	1	
7	0111	0	
8	1000	0	
9	1001	0	
10	1010	1	
11	1011	1	
12	1100	1	
13	1101	0	
14	1110	1	
15	1111	0	
16	0000	1	
17	0001	0	(The pattern begins to repeat here.)
18	0010	1	
19	0011	0	
20	0100	0	
21	0101	1	
22	0110	1	
23	0111	0	
24	1000	0	

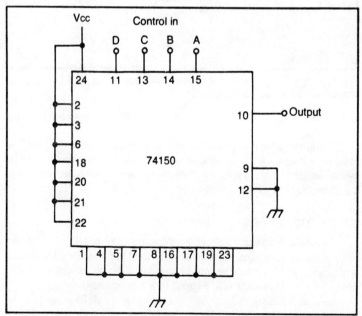

Fig. 15-4. This circuit is much simpler, but it accomplishes the same thing as the circuit shown in Fig. 15-2.

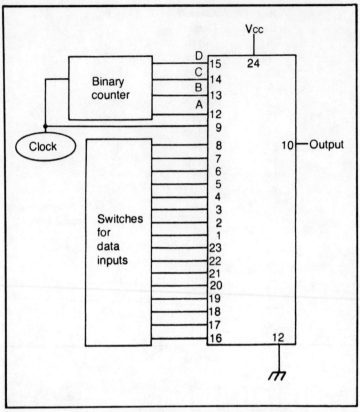

Fig. 15-5. Combining a 74150 1-of-16 multiplexer with a 4-bit modulo-16 counter produces a binary pattern generator.

and so on. Any pattern of binary digits can easily be generated by using this method.

Multiplexers have many additional applications. For instance, they are often used to scan a series of switchers or a keyboard, testing each individual switch (or key) to see whether it is open or closed, and telling the rest of the circuitry to respond accordingly.

DEMULTIPLEXERS

The opposite of a multiplexer is a demultiplexer, or DEMUX. Where the control inputs of a multiplexer determine which of several data inputs will be seen at the output, the control inputs on a demultiplexer determine which of several outputs will respond to a single data input.

A simple demultiplexer can be constructed from NAND gates, just like the multiplexer described earlier in this chapter. A simple 1-of-4 demultiplexer circuit is illustrated in Fig. 15-6. The action of this circuit can be

334

described with this truth table:

Data Input	Control Inputs		Outputs			
	A	B	1	2	3	4
0	0	0	0	1	1	1
1	0	0	1	1	1	1
0	0	1	1	0	1	1
1	0	1	1	1	1	1
0	1	0	1	1	0	1
1	1	0	1	1	1	1
0	1	1	1	1	1	0
1	1	1	1	1	1	1

Only one output is active at any time. Inactive outputs are held at logic 1 (HIGH). In some demultiplexer circuits, inactive outputs may be held at logic 0 (LOW).

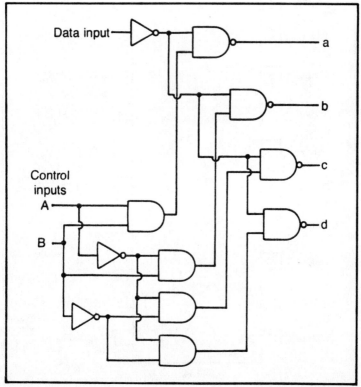

Fig. 15-6. Standard logic gates can also be used to create a demultiplexer circuit.

Figure 15-7 shows a popular 1-of-16 demultiplexer IC. This is the 74154. Like the 74150 multiplexer discussed in the preceding section, the 74154 belongs to the TTL logic family.

A demultiplexer is often employed to convert binary numbers into another number system. For example, the 74154 can convert four-digit binary (base two) numbers into single-digit hexadecimal (base 16) numbers with the circuit shown in Fig. 15-8. For any combination of control inputs (any binary number from 0000 to 1111), one, and only one of the outputs

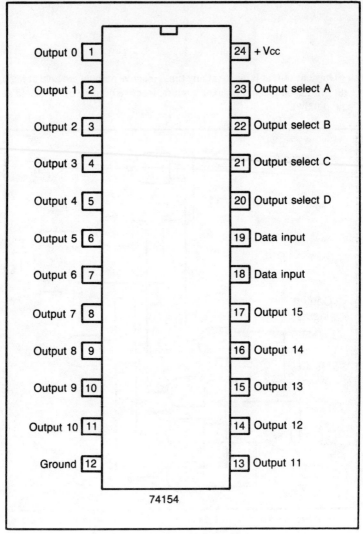

Fig. 15-7. The 74154 is a 1-of-16 demultiplexer IC.

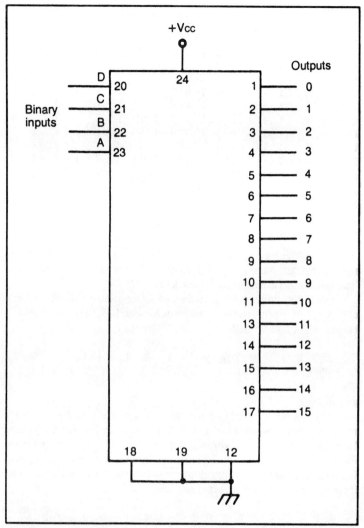

Fig. 15-8. A demultiplexer can be used as a binary-to-hexadecimal converter.

will be at logic 0. The other fifteen outputs will be held at logic 1:

Control Inputs				Outputs															
A	B	C	D	0	1	2	3	4	5	6	7	8	9	10	11	12	13	14	15
0	0	0	0	0	1	1	1	1	1	1	1	1	1	1	1	1	1	1	1
0	0	0	1	1	0	1	1	1	1	1	1	1	1	1	1	1	1	1	1

Control Inputs				Outputs															
0	0	1	0	1	1	0	1	1	1	1	1	1	1	1	1	1	1	1	1
0	0	1	1	1	1	1	0	1	1	1	1	1	1	1	1	1	1	1	1
0	1	0	0	1	1	1	1	0	1	1	1	1	1	1	1	1	1	1	1
0	1	0	1	1	1	1	1	1	0	1	1	1	1	1	1	1	1	1	1
0	1	1	0	1	1	1	1	1	1	0	1	1	1	1	1	1	1	1	1
0	1	1	1	1	1	1	1	1	1	1	0	1	1	1	1	1	1	1	1
1	0	0	0	1	1	1	1	1	1	1	1	0	1	1	1	1	1	1	1
1	0	0	1	1	1	1	1	1	1	1	1	1	0	1	1	1	1	1	1
1	0	1	0	1	1	1	1	1	1	1	1	1	1	0	1	1	1	1	1
1	0	1	1	1	1	1	1	1	1	1	1	1	1	1	0	1	1	1	1
1	1	0	0	1	1	1	1	1	1	1	1	1	1	1	1	0	1	1	1
1	1	0	1	1	1	1	1	1	1	1	1	1	1	1	1	1	0	1	1
1	1	1	0	1	1	1	1	1	1	1	1	1	1	1	1	1	1	0	1
1	1	1	1	1	1	1	1	1	1	1	1	1	1	1	1	1	1	1	0

Demultiplexers are also frequently used for decoding data. For this reason, they are often referred to as decoders. By the same token, multiplexers are occasionally called encoders.

If the four inputs of a 1-of-16 demultiplexer are fed with the outputs of a modulo-16 binary counter, as illustrated in Fig. 15-9, the demultiplexer's outputs will cycle through in sequence, like this:

0 1 2 3 4 5 6 7 8 9 10 11 12 13 14 15 0 1 2 3
4 5 6 7 8 9 10 11 12 13 14 15 0 1 2 3 4 5. . .

and so forth.

For even more versatility, we can take advantage of the CLEAR (or RESET) input that is available on most binary counter circuits. By feeding back the appropriate output from the demultiplexer to the counter CLEAR, the counting sequence can easily be limited to any desired maximum value. For instance, feeding back demultiplexer output eight to the counter CLEAR input would result in this output sequence:

0 1 2 3 4 5 6 7 0 1 2 3 4 5 6 7 0 1 2 3 4

and so on.

With some circuits an inverter may be needed on this feedback signal for correct operation. If the wrong logic signal is fed into the CLEAR input, the outputs will get stuck in a very nonuseful rut:

0 0 0 0 0 0. . .

and so on, through eternity. Clearly there wouldn't be much practical application for that.

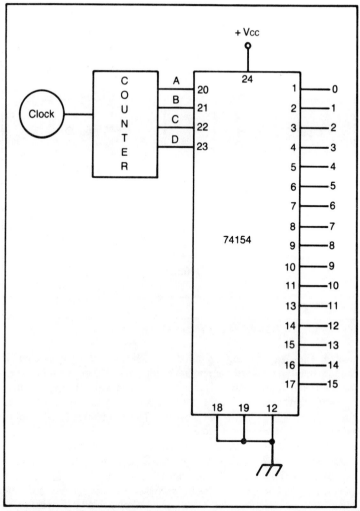

Fig. 15-9. Driving a demultiplexer with a counter causes the outputs to cycle through in sequence.

BILATERAL SWITCHES

We will conclude this chapter with a look at bilateral switches. Bilateral switches are not really related to multiplexers and demultiplexers, but they are often used in more or less similar applications, so I am including them in this chapter. They don't really fit in anywhere else either.

A bilateral switch is nothing more than a digitally controllable switch. A logic 0 opens the switch, and a logic 1 closes it. The switch itself can be used in any switching application, just as if it was a regular mechanical

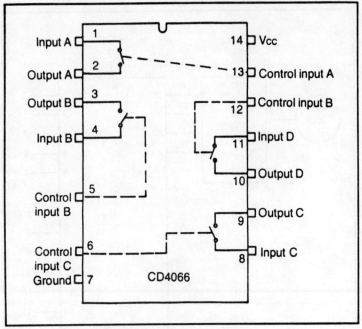

Fig. 15-10. The CD4066 is a quad bilateral switch IC.

switch. The signal passing through the switch can be almost anything (as long as it doesn't exceed the maximum ratings of the device, of course). Either digital or analog signals may be controlled with a bilateral switch.

The term "bilateral" simply means that the signal may pass through the switching element in either direction. There is no dedicated input or output.

Figure 15-10 shows the CD4066 quad bilateral switch IC. This CMOS device contains four independent bilateral switches, each with its own digital control input. The switch element contacts are labeled as inputs and outputs strictly for convenience. The inputs and outputs may be reversed without affecting circuit operation. (The control inputs are digital inputs only.)

Digital switch units like the CD4066 are often used in hybrid circuits in which both analog and digital devices are used. Analog components, such as resistors and capacitors, may be selected or programmed via digital control signals. Digital-to-analog signal conversion is one of the many possible applications for this handy device.

340

```
     ┌──┐ ∪ ┌──┐
   ┌─┤ 1│   │24├─┐
   ┌─┤ 2│ S │23├─┐
   ┌─┤ 3│ P │22├─┐
   ┌─┤ 4│ E │21├─┐
   ┌─┤ 5│ C │20├─┐
   ┌─┤ 6│ I │19├─┐
   ┌─┤ 7│ A │18├─┐
   ┌─┤ 8│ L │17├─┐
   ┌─┤ 9│ P │16├─┐
   ┌─┤10│ U │15├─┐
   ┌─┤11│ R │14├─┐
   ┌─┤12│ P │13├─┐
            O
            S
            E
```

Chapter 16

Some Unusual Digital Devices

T HE DIGITAL ICS WE HAVE DISCUSSED SO FAR WERE ALL DESIGNED
to perform specific, but fairly common functions. Some special-
purpose ICs are more specialized than others. In this chapter we will look
at a few of the less frequently encountered digital chips.

THE 14415 TIMER/DRIVER

Back in Chapter 3 we examined ICs intended for use in analog timing
applications. We mentioned that most of these devices can also be interfaced with digital circuitry. The 14415, which is shown in Fig. 16-1 is a
true digital timer chip. A block diagram of this device's internal circuitry
is shown in Fig. 16-2.

This CMOS IC actually contains four timer stages. The pulse width
of each timer's output is a function of the input clock frequency. When
the proper input sequence is applied to the chip, one of the output buffers
is turned on (set). Now the 14415 starts to count the incoming clock pulses.
After a count of 100, the output buffer is turned back off (reset).

The 14415 was designed primarily for use in high speed line printers.
The hammer drivers in such printers demand very precise timing for proper
operation. This chip can also be used in any application requiring critical
timing and precise pulse widths.

THE 14410 TONE ENCODER

The 14410 two-of-eight tone encoder IC, which is shown in Fig. 16-3,
could have been included in Chapter 11 (Telephone ICs) along with the
TCM5089 tone encoder IC. As with the TCM5089, the primary applica-

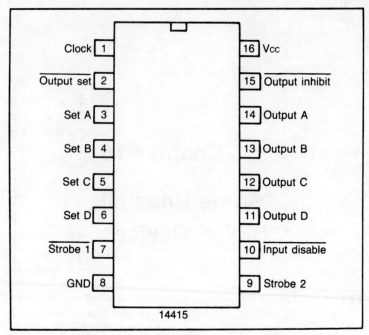

Fig. 16-1. The 14415 is a true digital timer IC.

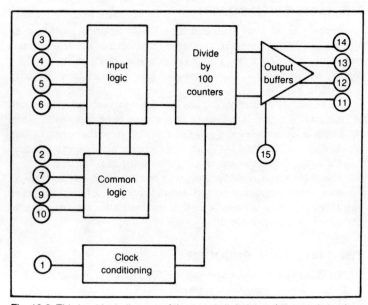

Fig. 16-2. This is a block diagram of the 14415 digital timer's internal circuitry.

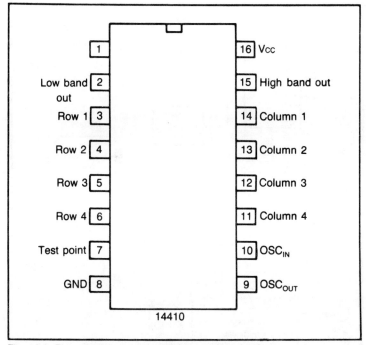

Fig. 16-3. The 14410 is a two-of-eight tone decoder IC.

tion for the CMOS 14410 is to synthesize the high band and low band sine-wave tones utilized in the telephone Touch Tone® system.

The 14410 accepts digital inputs in a two-of-eight code format, and digitally synthesizes the appropriate tones. An on-chip crystal-controlled oscillator is used as the master clock signal source for the chip. Of course, the crystal itself must be added externally.

The inputs are normally arranged in the standard 4 × 4 row/column matrix format. Whenever a key is depressed, one row and one column are activated (connected to Vss). Internal clocks controlling the logic are enabled only by one or more row and column being activated simultaneously.

The output stages include NPN bipolar transistors. This allows the 14410 to handle drive low-impedance outputs, and to source relatively large currents. A block diagram of the 14410's internal circuitry is shown in Fig. 16-4.

KEY ENCODERS

Many digital applications require data to be entered via some form of keyboard, or array of switches. There must be some way to encode the entered data into a form recognizable by the digital circuitry. Special-purpose ICs for this function are called key encoders. A typical key en-

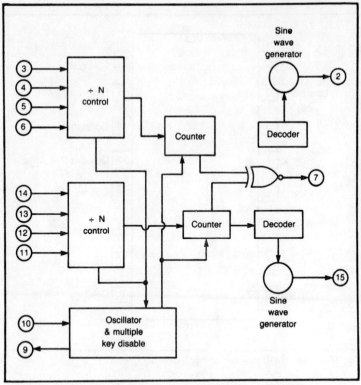

Fig. 16-4. This is a block diagram of the 14410's internal circuitry.

coder IC is the CMOS 74C922, which is shown in Fig. 16-5. This chip contains all the necessary logic to fully encode a 16-key array of SPST switches.

Each of the keys is scanned sequentially under the control of an internal or an external oscillator. The frequency of the internal oscillator is determined by an external capacitor. The 74C922 features on-chip pull-up resistors. Switches with up to 50K of resistance may be used with this device.

Many keyboard encoding circuits require an array of diodes to prevent phantom switching—that is, the circuit may get confused about which switch is depressed due to crosstalk between switches. No such diodes are required by the 74C922.

This IC also features internal debounce circuitry. Mechanical switches tend to bounce on and off several times very rapidly when first depressed or released. For most applications this doesn't really matter very much, but many digital circuits are fast enough to count each bounce as a separate switch closure. A debouncing circuit adds a slight delay, so the digital circuitry isn't so easily confused by the mechanical vibration of the switch. A single external capacitor is needed to give the 74C922 debounce protection. Alternatively, the debounce circuitry can be defeated, simply by omitting this external capacitor.

Pin 12 is labelled DATA ENABLE. This pin goes HIGH when a valid keyboard entry is made. It can be used to synchronize the 74C922 with other circuitry, or to activate interrupts. Two-key rollover is provided between any two switches. This speeds up data entry. An internal register remembers which key was depressed last, even after the switch has been released.

The outputs of the 74C922 use three-state logic (see Chapter 14). The 74C922 is designed to encode a 16-key keyboard. A closely related device is the 74C923, which can encode up to 20 keys. This device is illustrated in Fig. 16-6.

SWITCH DEBOUNCER

No switch is ever perfect. They are all subject to some mechanical vibration when they are opened and closed. This is called switch bounce. For most applications this doesn't really matter very much, but many digital circuits are fast enough to count each bounce as a separate switch closure.

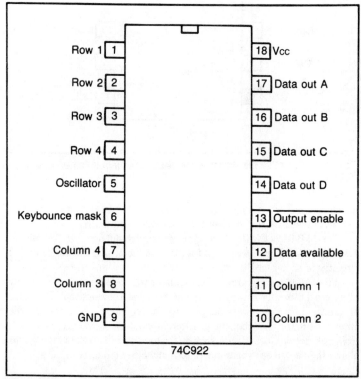

Fig. 16-5. The 74C922 contains all the necessary logic to fully encode a 16-key array of SPST switches.

345

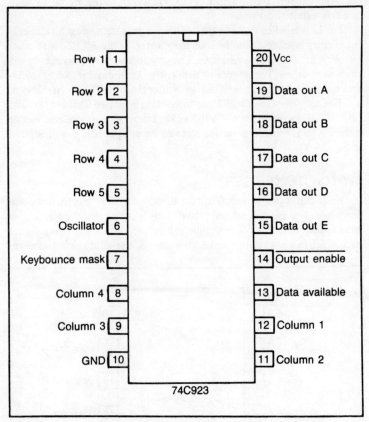

	74C923	
Row 1 ⌷1		20⌷ Vcc
Row 2 ⌷2		19⌷ Data out A
Row 3 ⌷3		18⌷ Data out B
Row 4 ⌷4		17⌷ Data out C
Row 5 ⌷5		16⌷ Data out D
Oscillator ⌷6		15⌷ Data out E
Keybounce mask ⌷7		14⌷ Output enable
Column 4 ⌷8		13⌷ Data available
Column 3 ⌷9		12⌷ Column 1
GND ⌷10		11⌷ Column 2

Fig. 16-6. The 74C923 is similar to the 74C922, except it can encode up to 20 keys.

A debouncing circuit adds a slight delay, so the digital circuitry isn't so easily confused by the mechanical vibration of the switch.

The 14490 is a CMOS IC designed for switch debouncing. The pinout diagram of this device is given in Fig. 16-7. A block diagram of its internal circuitry is shown in Fig. 16-8.

The 14490 has an internal RC oscillator which serves as a clock. Only an external capacitor is needed to make this oscillator functional. The value of the capacitor determines the oscillator's frequency, and therefore, the amount of delay for debouncing. Alternatively, the clock may be driven from an external clock source, such as the oscillator of another 14490. This allows the chip to be synchronized with any other circuitry in the system.

RATE GENERATORS/MULTIPLIERS

Figure 16-9 shows the 14411 CMOS bit-rate generator IC. A block di-

346

agram of this device is shown in Fig. 16-10. This chip's circuitry is built around a frequency-divider network, that allows it to generate a wide range of output frequencies. An external crystal is used to set the basic clock frequency. The nominal frequency is 1.8432 MHz.

A 2-bit digital value is fed to pins 22 and 23 to select one of four multiple output clock rates. Sixteen different output clock rates are available. The output signals all have a 50% duty cycle (square wave). This device has many applications as a selectable frequency source for data communications equipment, including:

- □ CRT terminals
- □ Microprocessor systems
- □ Printers
- □ Teletypes

A similar device is the 4089 binary-rate multiplier, which is shown in Fig. 16-11. This chip accepts a clock input signal, and a 4-bit binary number input. The output pulse rate is one-sixteenth of the input clock frequency multiplied by the binary data input. In other words, the output

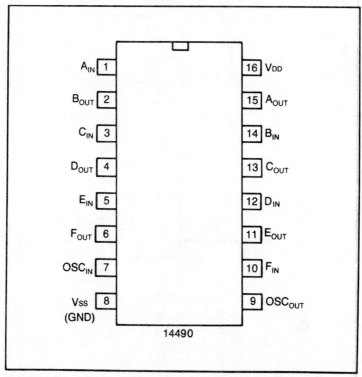

Fig. 16-7. The 14490 is a CMOS IC designed for switch debouncing.

347

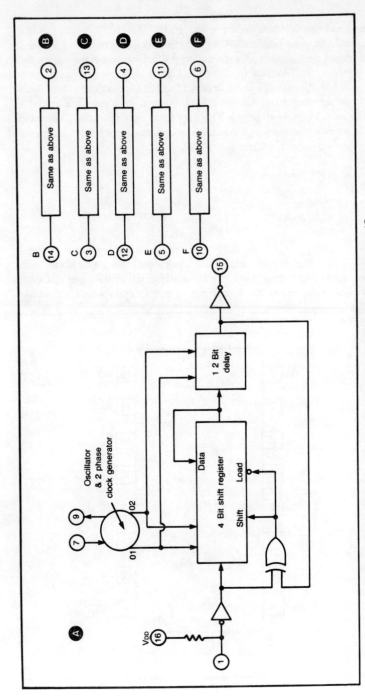

Fig. 16-8. This block diagram illustrates the internal circuitry of the 14490 switch debouncer IC.

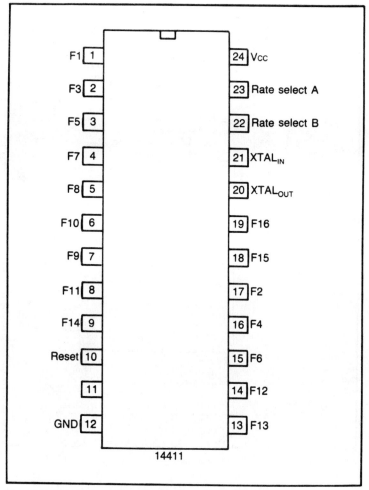

Fig. 16-9. This is the 14411 CMOS bit-rate generator IC.

frequency can be defined with this general formula:

$$F_o = (F_c/16) \times b$$

where F_o is the output frequency, F_c is the input clock frequency, and b is the decimal equivalent of the binary input data.

As an example, let's say the clock frequency is 3.2 MHz, and the binary number being input is 0110 (decimal 6). The output would be equal to:

$$F_o = (3,200,000/16) \times 6$$
$$= 200,000 \times 6$$

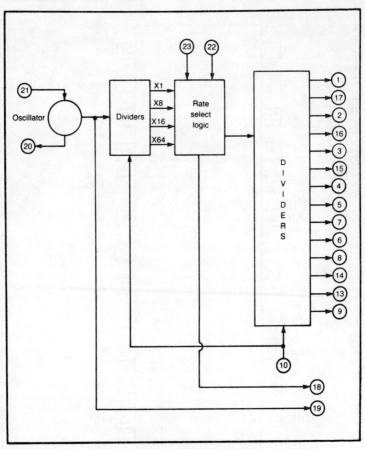

Fig. 16-10. This is a block diagram of the 14411's internal circuitry.

$$= 1{,}200{,}000 \text{ Hz}$$
$$= 1.2 \text{ MHz}$$

Another way to express the output frequency in this example is that there are six output pulses for every sixteen input pulses.

The 4089 can be used to perform arithmetic operations such as multiplication and division. This device can also be employed in A/D (analog-to-digital) and D/A (digital-to-analog) conversion. Obviously, frequency division is another good application for the 4089 binary bit multiplier.

Because of the divide-by-16 constant, the 4089 is ideal for binary and hexadecimal oriented applications, but it would obviously be awkward for decimal applications. When working with decimal numbers, a better choice would be the 4527 BCD rate multiplier. In this chip the data is in BCD (binary-coded-decimal) form. The clock frequency is divided by 10 (instead of 16) before being multiplied by the input data. The formula is:

$$F_o = (F_c/10) \times b$$

The terms have the same meanings used in the formula for the 4089. The 4527 BCD rate multiplier IC is illustrated in Fig. 16-12.

PARITY GENERATORS/CHECKERS

In transmitting digital data, it is critical that the received signal be absolutely error free. One wrong bit can result in a string of utter garbage, rather than meaningful data. But the receiving system may try to use the garbage as if it was meaningful data. How can it tell the difference?

One method used in some form in most data transmission schemes is parity. There are several different variations on this idea in use. We will just discuss the most basic approach.

One bit of each transmitted digital word (multi-digit binary number) is reserved for a parity value. It is not actually part of the data. Let's say we have a system that transmits data in 8-bit chunks (bytes). Seven of the bits will carry the actual data. The eighth is a parity bit. The value of the

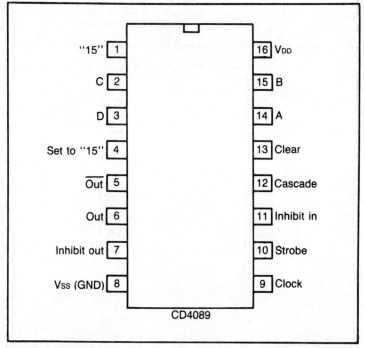

Fig. 16-11. The 4089 binary rate multiplier is related to the 14411 bit-rate generator of Fig. 16-9.

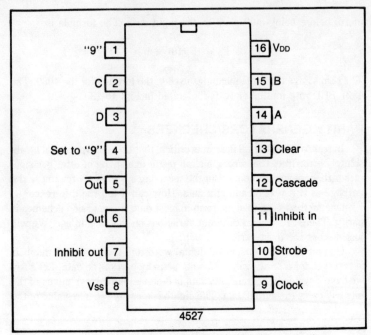

Fig. 16-12. The 4527 rate-multiplier IC is used for BCD applications.

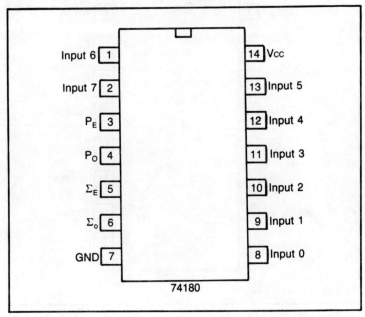

Fig. 16-13. The 74180 generates and/or checks for correct parity bit.

352

parity bit is determined by the other seven bits. If the data bits add up to an odd value, the parity bit is made a 1. If the data bits add up to an even value, the parity bit is a 0.

When the receiving system looks at an incoming byte, it adds up the digits and compares its results with the received parity bit. If the received and the locally calculated parity bit agree, the data is assumed to be accurate, and is accepted. Otherwise, it is rejected.

Of course, it is theoretically possible that two complementary errors could conceivably occur in a single byte and make the parity bit look correct, even when the data is erroneous. But it is reasonable to assume that the odds of this happening are slight.

The 74180 is a 9-bit parity generator/checker. This TTL chip is illustrated in Fig. 16-13. It adds a ninth parity bit to 8-bit data. Either odd or even parity may be used.

Chapter 17

CPUs and

Related Devices

A S INTEGRATED CIRCUIT TECHNOLOGY HAS IMPROVED OVER THE years, more and more special-purpose ICs have appeared. Generally, the more complex the internal circuitry, the more specific the function of the IC is. A relatively simple operational amplifier (op amp) has literally hundreds of potential applications. The melody generators discussed in Chapter 9 are dedicated devices. They do one complex task very well, and little, if anything, else.

Curiously though, one of the most complex types of special purpose ICs has countless applications. Its possibilities are so wide, that it has become trite to declare it has ushered in a revolution and a new age. Trite, but true. Of course, the special-purpose IC I am speaking of is the CPU (central processing unit). It is sometimes called a computer on a chip, which isn't entirely accurate yet, but the industry is getting closer to this point.

Advanced CPU ICs are the equivalent of thousands of discrete components. It is a very complex and dense type of circuitry, designed for a very specific purpose. But that specific purpose includes programmability—the chip "understands" dozens of instruction codes that can be combined for millions of different uses. It is ironic that the most specialized of special-purpose ICs offers almost unlimited versatility.

THE 74181 ALU

Before looking at some actual CPU ICs, let's first consider a somewhat simpler device. The part of a CPU that does the actual mathematical operations is the *arithmetic logic unit*, or ALU. An independent 4-bit ALU IC is the 74181, which is shown in Fig. 17-1. The following is a capsule

354

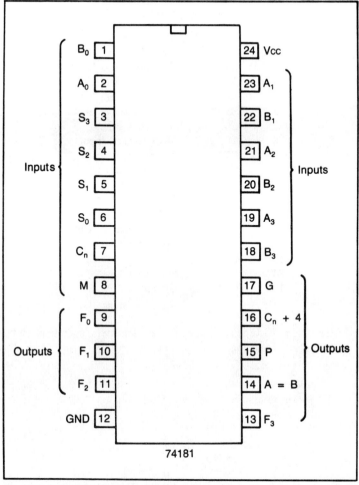

Fig. 17-1. The 74181 ALU can perform various arithmetic and logic operations on two 4-bit binary numbers.

summary of the 74181's pins, and their functions:

Pin	Label	Function
2	A0	Word A Input
23	A1	Word A Input
21	A2	Word A Input
19	A3	Word A Input
1	B0	Word B Input
22	B1	Word B Input

Pin	Label	Function
20	B2	Word B Input
18	B3	Word B Input
6	S0	Function Select Input
5	S1	Function Select Input
4	S2	Function Select Input
3	S3	Function Select Input
7	Cn	Inv. Carry Input
8	M	Mode Control Input
9	F0	Function Output
10	F1	Function Output
11	F2	Function Output
13	F3	Function Output
14	A = B	Comparator Output
15	P	Carry Propagate Output
16	Cn + 4	Inv. Carry Output
17	G	Carry Generate Output
24	Vcc	Supply Voltage
12	Gnd	Ground

The 74181 ALU can perform 16 binary arithmetic operations on two 4-bit numbers, including:

☐ Addition
☐ Subtraction
☐ Decrement
☐ Shift
☐ Straight transfer
☐ Magnitude comparison

and others.

The operation to be performed is selected by the data applied to the four-function input (S0 through Sa—pins 3-6). Internal carries for the various arithmetic operations must be enabled by applying a LOW signal to the mode control input (M—pin 8).

High speed arithmetic operations can be performed by the 74181 by combining it with a *look ahead carry generator*, like the 74182, which is shown in Fig. 17-2. The way these two devices should be wired together is illustrated in Fig. 17-3. The carry ahead scheme is supported by means of two cascaded outputs (P—pin 15 and G—pin 17) for the four bits in the package.

The following is a summary of the arithmetic/logic operations performed for each of the select options (as applied to pins 6 - 3). The data is assumed to be active HIGH (L = 0, H = 1):

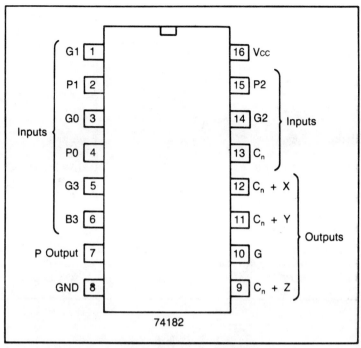

Fig. 17-2. This is the 74182 look-ahead-carry generator.

Select	M = H	M = L	Arithmetic Operations
	Logic Functions	C_n = H (no carry)	C_n = L (with carry)
LLLL	$F = \overline{A}$	$F = A$	$F = A$ plus 1
LLLH	$F = \overline{A + B}$	$F = A + B$	$F = (A + B)$ plus 1
LLHL	$F = \overline{A}B$	$F = A + \overline{B}$	$F = (A + \overline{B})$ plus 1
LLHH	$F = 0$	F = minus 1 (TWOs complement)	F = zero
LHLL	$F = \overline{AB}$	$F = A$ plus $A\overline{B}$	$F = A$ plus $A\overline{B}$ plus 1
LHLH	$F = \overline{B}$	$F = (A + B)$ plus $A\overline{B}$	$F = (A + B)$ plus $A\overline{B}$ plus 1
LHHL	$F = A \oplus B$	$F = A$ minus B minus 1	$F = A$ minus B
LHHH	$F = A\overline{B}$	$F = A\overline{B}$ minus 1	$F = A\overline{B}$
HLLL	$F = \overline{A} + B$	$F = A$ plus AB	$F = A$ plus AB plus 1

357

Select	M = H	M = L	Arithmetic Operations
HLLH	$F = \overline{A \oplus B}$	F = A plus B	F = A plus B plus 1
HLHL	$F = \overline{B}$	$F = (A + \overline{B})$ plus AB	$F = (A + \overline{B})$ plus AB plus 1
HLHH	$F = A\overline{B}$	F = AB minus 1	$F = A\overline{B}$
HHLL	F = 1	F = A plus A*	F = A plus A plus 1
HHLH	$F = A + \overline{B}$	F = (A + B) plus A	F = (A + B) plus A plus 1
HHHL	F = A + B	$F = (A + \overline{B})$ plus A	$F = (A + \overline{B})$ plus A plus 1
HHHH	F = A	F = A minus 1	F = A

(*each bit is shifted to the next more significant position)   (AB = A AND B)   (A + B = A OR B)   ($A \oplus B$ = A X-OR B)   ($\overline{A}$ = NOT A)   ($\overline{B}$ = NOT B)   ($\overline{AB}$ = A NAND B)   ($\overline{A + B}$ = A NOR B)

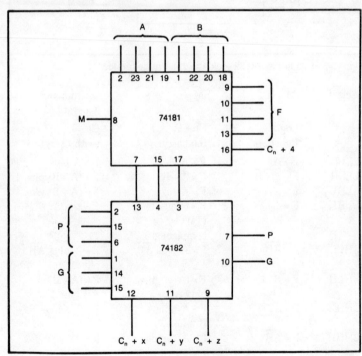

Fig. 17-3. High speed arithmetic operations can be performed by combining a 74181 ALU with a 74182 look-ahead-carry generator.

As you can see, the 74181 can perform a lot of different operations, but that's not all. By using active LOW logic (L = 1 and H = 0), we get the following results:

Select	M = H	M = L		Arithmetic Operations
	Logic Functions	$C_n = H$ (no carry)	$C_n = L$ (with carry)	
LLLL	$F = \overline{A}$	$F = A$ minus 1	$F = A$	
LLLH	$F = \overline{AB}$	$F = AB$ minus 1	$F = AB$	
LLHL	$F = \overline{A} + B$	$F = \overline{AB}$ minus	$F = A\overline{B}$	
LLHH	$F = 1$	$F = $ minus 1 (TWOs complement)	$F = $ zero	
LHLL	$F = \overline{A + B}$	$F = A$ plus $(A + \overline{B})$	$F = A$ plus $(A + \overline{B})$ plus 1	
LHLH	$F = \overline{B}$	$F = AB$ plus $(A + \overline{B})$	$F = AB$ plus $(A + \overline{B})$ plus 1	
LHHL	$F = \overline{A \oplus B}$	$F = A$ minus B minus 1	$F = A$ minus B	
LHHH	$F = A + \overline{B}$	$F = A + \overline{B}$	$F = (A + \overline{B})$ plus 1	
HLLL	$F = \overline{A}B$	$F = A$ plus $(A + B)$	$F = A$ plus $(A + B)$ plus 1	
HLLH	$F = A \oplus B$	$F = A$ plus B	$F = A$ plus B plus 1	
HLHL	$F = B$	$F = A\overline{B}$ plus $(A + B)$	$F = A\overline{B}$ plus $(A + B)$ plus 1	
HLHH	$F = A + B$	$F = A + B$	$F = (A + B)$ plus 1	
HHLL	$F = 0$	$F = A$ plus A*	$F = A$ plus A plus 1	
HHLH	$F = A\overline{B}$	$F = AB$ plus A	$F = AB$ plus A plus 1	
HHHL	$F = AB$	$F = A\overline{B}$ plus A	$F = A\overline{B}$ plus A plus 1	
HHHH	$F = A$	$F = A$	$F = A$ plus 1	

In some applications, the high speed operations may not be required. In such cases the ripple carry input (C_n—pin 7) and output ($C_n + 4$—pin 16) may be used. Actually, the ripple carry delay has been minimized, so that arithmetic manipulation of small numbers can be performed without external circuitry.

The 74181 can also perform comparator functions. The A = B output (pin 14) goes HIGH when all four F outputs are HIGH in the subtract mode, indicating that the two input values (A and B) are of equal magnitude. By using the A = B with the Cn + 4 ripple carry output, indications of A > B and A < B may also be achieved.

FOUR BIT CPUs

The first CPU ICs were 4-bit devices. That is, each operation and address must be made up of 4-bit "nibbles." While offering considerably more power than any IC prior to their development, the early 4-bit CPUs left quite a bit to be desired when it came to speed and memory addressing capabilities.

A typical 4-bit microprocessor IC is Intel's 4004. This chip could understand and act out 46 instructions, with both binary and decimal arithmetic being supported. The 4004 can directly address up to 8 K (8 × 1024) of ROM (read only memory) and 1 K of RAM (random access, or read-write, memory). This memory limitation stands in the way of serious programming for the most part.

A 4-bit CPU like the 4004 is best suited for dedicated applications where it is expected to perform a few specific tasks, and user programmability is of little interest. Typical applications include microwave ovens, and smart automobile metering.

THE 8080 CPU

Probably the first "best-seller" CPU was the 8080, also from Intel. While obsolete in many ways, the 8080 is still available, and can be used in a great many applications. This IC is a full 8-bit microprocessor. Actually, the 8080 is an upgraded version of the earlier and simpler 8008. The 8080 uses 16-bit addressing, so it can directly access up to 64 K (64 × 1024) of memory, or up to 256 I/0 (input/output) devices.

External stacking is used by the 8080. This means that external memory locations are used to store stack operations, rather than internal registers. This allows the stacks to be of (theoretically) infinite size. Three power-supply voltages are required for the 8080 to function. They are as follows:

+ 12 volts
+ 5 volts
-5 volts

All three supply voltages must be fairly precise, and well filtered.

The 8080 can operate at clock speeds from 500 kHz (500,000 Hz) to 2 MHz (2,000,000 Hz). The clock signal must be generated by external circuitry.

While it was initially billed as a "computer on a chip," the 8080 most definitely is not a complete microcomputer by itself. In addition to the external memory devices (discussed later in this chapter), a number of external support chips are required for operation of the 8080. The 8080 is illustrated in Fig. 17-4.

THE Z80 CPU

An even more advanced CPU is the Z80, which was designed by Zilog. This powerful IC is illustrated in Fig. 17-5. This 8-bit microprocessor was designed to be downward compatible with the 8080. This means that a machine-language program written on an 8080-based microcomputer will run directly on a Z80 based machine. The reverse is not necessarily true, however, since the Z80's instruction set includes a number of commands that are not available on the 8080.

Some versions of the Z80 can operate off an externally generated clock frequency of up to 4 MHz (4,000,000 Hz). This means that in some cases a Z80 can function twice as fast as an 8080.

THE 8085 CPU

Another improved version of the 8080 is the 8085, also by Intel. This CPU chip is illustrated in Fig. 17-6. While not pin compatible with the 8080, the 8085 is downwards software compatible with its popular predecessor. That is, a program written on an 8080-based computer can run directly on an 8085 based machine. The reverse may or may not be true, because the 8085's instruction set includes a few commands that are not recognized by the 8080.

One of the biggest disadvantages of the 8080 was the number of external support chips were required for a functional computer system. Dozens of TTL ICs were required for an 8080-based microcomputer. On the 8085 many of the functions performed by the external support chips for the 8080 are performed on chip by the 8085 itself. This greatly simplifies circuit design. For example, the 8080 requires an external two-phase clock to operate. The 8085 includes an on-chip clock oscillator. Only an external crystal, or RC network are required to set the clock frequency. The nominal clock frequency for the 8085 is 3 MHz (3,000,000 Hz).

The 8085 runs faster than the 8080. The basic instruction cycle for the 8080 is 2 μs (0.000002 second) long, while the basic instruction cycle for the 8085 is only 1.3 μs (0.0000013 second) long. A difference of 0.7 millionths of a second may not seem like much, but when you multiply it by the millions of instructions executed in a typical program, the cumulative difference can become quite significant.

THE 6800 CPU

Of course, not all 8-bit CPU ICs are direct variations on the 8080. Many other manufacturers have designed their own CPUs more or less from the ground up. Another popular family of CPU ICs are the 6800 and related

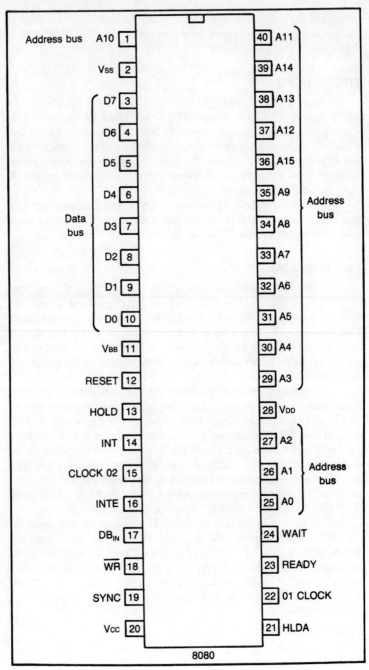

Fig. 17-4. The 8080 was the first "best-seller" CPU IC.

362

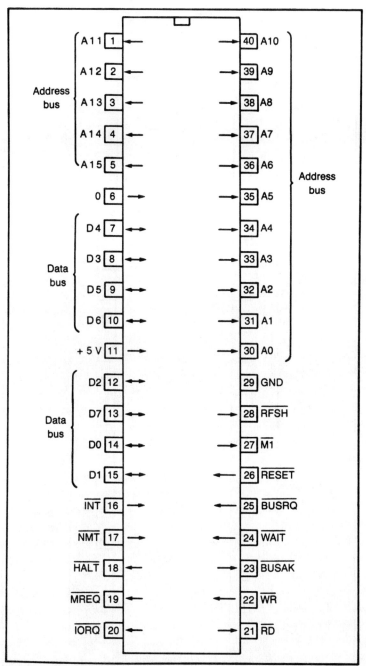

Fig. 17-5. The Z80 is essentially an improved version of the 8080.

363

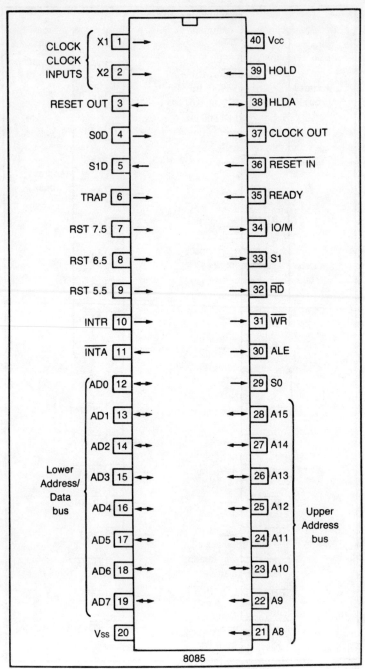

Fig. 17-6. Another improved version of the 8080 is the 8085.

devices. The 6800 was originally developed by Motorola. This chip's instruction set includes 72 commands. An off-chip clock is required for operation.

As with the 8080, the 6800 has spawned a number of more powerful CPUs that are downward compatible with the original unit. The 6802 is an improved version of the 6800. Its new features include an on-chip clock oscillator and 128 bytes of internal RAM (random access memory).

Another variation on the 6800 is the 6809, which uses a 16-bit architecture. Software written on a 6800-based machine are not directly usable on the 6809, but they can easily be reassembled to run on this more powerful microprocessor.

THE MCS-48 CPUs

Often a CPU IC is called a computer on a chip. Usually this isn't really true, since a number of external components are required to create a functional microcomputer system. Devices in the MCS-48 series, however, could be considered virtually complete by themselves.

The 8048 is the most deluxe model. Besides a full 8-bit CPU, this chip includes 1 K of ROM (read only memory), 64 bytes of RAM (random access, or read/write memory), an 8-bit timer/event counter, and 27 I/O (input/output lines). All of this is contained in a single IC package. Now, that's a computer on a chip. The internal memory is rather limited, of course, but it is adequate for many dedicated applications (in video equipment, microwave ovens, automobiles, etc.). External memory devices may also be added to the system if desired.

The 8748 is the same as the 8048, except the 1 K or ROM is actually EPROM (erasable programmable read only memory), permitting greater versatility and customization to the user's specific application(s).

An on-chip clock oscillator is included, only an external crystal, or LC network is required. Alternatively, an external clock signal source may be used. This chip operates on a single-ended + 5-volt supply. The 8748 is illustrated in Fig. 17-7.

SIXTEEN BIT CPUs

The 8080, Z80, and 6800 are all 8-bit CPUs. At one time this was the standard for microcomputers. Recently, however, 16-bit CPUs have been appearing on the market. In an 8-bit processor, the CPU registers and data paths are 8-bits wide. That is, instructions and data can only be operated on in 8-bit chunks. Larger units may be used, but they would require more instruction cycles. For example, a 16-bit value would have to be broken up into two 8-bit chunks—each accessed and operated on separately.

Some instructions for 8-bit CPUs are 16-, or even 24-bits long. They must be broken up into two or three 8-bit bytes. This is awkward and slow. A 16-bit CPU, however, features 16-bit registers and data paths. There-

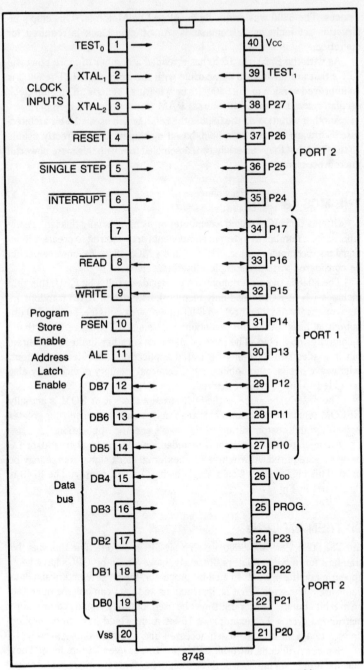

Fig. 17-7. The 8748 can be considered a true computer-on-a-chip.

366

fore, it can deal directly with 16-bit instructions and data words of sixteen bits. In addition, a much larger amount of memory can be directly accessed.

Eight-bit and 16-bit CPUs have instruction cycle times that are more or less equal. Since the 16-bit unit can take in twice as much data each cycle, it naturally follows that it operates about twice as fast as a comparable 8-bit device.

Sixteen-bit CPUs can also deal with multi-word instructions, just as the 8-bit microprocessor can handle multi-byte instructions. In a 16-bit CPU, 32-bit and 48-bit instructions are broken up into 16-bit chunks that are handled individually. These long instructions can permit some very sophisticated operations.

An 8-bit CPU can perform all the functions performed by a 16-bit CPU, but it may be considerably slower and more awkward to program. For instance, this is a typical program segment for an 8-bit microprocessor. This program segment executes the BASIC statement A = B(7):

```
LXI     H,B
LXI     D,7
DAD     D
MOV     E,M
XCHG
SHLD    A
```

For a 16-bit CPU, the same function can be achieved with a much simpler program:

```
MOV     #7,R1
MOV     B(R1),A
```

A 16-bit CPU can be much faster and more efficient than an 8-bit CPU. However, in many applications the advantages will be slight, so 8-bit CPUs are certainly not obsolete yet.

Sixteen-bit microprocessors tend to be more expensive than 8-bit units. In addition, they require more expensive and elaborate support circuitry. Twice as many bus lines are required. Another disadvantage of 16-bit CPUs is that they tend to be rather inefficient when it comes to memory. A typical program might require 10% to 20% more memory space for a 16-bit system than for a comparable 8-bit system. But, since the 16-bit CPU can access much more memory than an 8-bit CPU, and memory costs have been droping so far, this memory inefficiency often isn't a problem. So there is a trade-off between 8-bit and 16-bit machines. The advantages of eight bit CPUs are:

☐ Less expensive
☐ Simpler support circuitry
☐ More memory efficient

The advantages of sixteen bit CPUs are:

☐ Faster
☐ More memory can be directly accessed
☐ More powerful instructions
☐ Can use some minicomputer programs and peripherals

Several 16-bit CPU ICs have appeared on the market in the past few years. These include:

☐ The Intel 8086
☐ The Zilog Z8000
☐ The NC68000

The 68000 is made by several manufacturers, including Mostek, Motorola, and Signetics.

Some devices are occasionally called 32-bit systems or 16/32-bit systems in some advertising literature. The 68000 is often advertised this way. It is a little misleading. Actually the 68000 is a 16-bit CPU, because its data bus is 16-bits wide. However, all of its internal registers are 32-bits wide, so in some ways it functions as if it was a 32-bit CPU.

In this chapter we have very briefly glanced at a few CPU ICs. Due to space limitations, we couldn't go into very much detail. Any one of these devices could easily warrant a book of its own. Many such books on specific CPUs have been written. My goal here was just to give you a passing familiarity with the basic differences between some of the important CPU ICs currently available.

Chapter 18

Memory Devices

I N MANY DIGITAL ELECTRONICS APPLICATIONS IT IS ESSENTIAL TO have some way to store digital data (binary numbers). This requirement is particularly pressing in computers and related equipment. If the amount of data to be stored is relatively small, the task can be accomplished with fairly simple digital circuits. A flip-flop, for example, can store a single bit. A longer binary word (string of bits) can be temporarily stored in a shift register.

Unfortunately, these techniques are extremely limited. Unless the amount of data to be stored is very small, flip-flops and shift registers are likely to prove completely inadequate for the job. Fortunately, special-purpose ICs have been designed for just this type of application. These devices are called memories. Many different types and sizes of digital memories are available in IC form.

RAM

One of the basic types of memory is called RAM. This is an abbreviation for random access memory. What this name means is that any specific location within the memory can be addressed without stepping through any of the other memory locations. A shift register, on the other hand, is an example of a sequential-access memory device. For instance, let's say we need to know the logic state of bit 5 in an 8-bit SISO (serial-in serial-out) shift register. Before we can reach bit 5, we have to step through bits 1, 2, 3, and 4. Clearly, this will take some finite amount of time. When dealing with thousands or millions of bits of data, the disadvantages of a sequential address memory system become painfully obvious.

	A	B	C	D
1	A1	B1	C1	D1
2	A2	B2	C2	D2
3	A3	B3	C3	D3
4	A4	B4	C4	D4
5	A5	B5	C5	D5

Fig. 18-1. Memory addressing is like a post office box system.

A random access memory, however, assigns a unique address to each individual storage location, with each location holding one piece of data (either a bit or a binary word). This type of memory addressing can be considered as an analogy to a post office box system. Any individual box (or memory location) can be uniquely defined by a specific address identifying its column and row. This idea is illustrated in Fig. 18-1.

Any specific piece of data can easily be found by going directly to its unique address. Rather than having to look at each storage location in sequence (1 - 2 - 3 - 4 - 5 - 6 - 7 -, etc.), we can go straight to the desired address — for instance, C7.

In a RAM we can either look at the value previously stored at a given address without changing it, or we can replace the old value with a new value. In other words, we can either read old data, or write new data. For this reason, RAM is sometimes called *read/write memory*, or *RWM*. Some technicians feel this is a better name, since ROM (discussed in the next section of this chapter) can also be randomly accessed. But RAM is the established name in common usage. There are two basic types of RAM. They are static RAM and dynamic RAM.

Static RAM

A static RAM is basically made up of a series of addressable flip-flops. Data can be stored in a static RAM virtually indefinitely. Unless the stored values are changed or erased, or the power supply is interrupted, the data remains unaltered. A static RAM cannot store data without continuously applied power, however. If the supply voltage is briefly interrupted for any

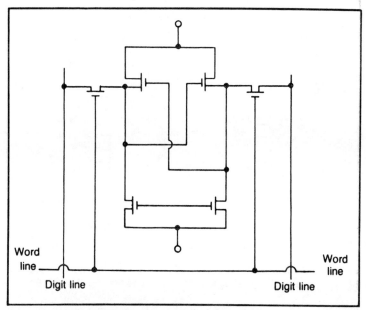

Fig. 18-2. A static RAM cell is basically a flip-flop stage.

reason whatsoever, the memory will "forget" everything it ever "knew. A typical static memory cell using CMOS technology is illustrated in Fig. 18-2.

One of the earliest static RAM ICs was Intel's MM1101, which was first made available in the late 1960s. This chip is shown in Fig. 18-3. Eight pins are used as a parallel bus for the memory address (A0 through A7). This means that the usable memory addresses may range from binary 0000 0000 (decimal 0) to 1111 1111 (decimal 255). In other words, the MM1101 static RAM IC contains 256 independent memory locations, each of which can store the value of a single bit (0 or 1).

Larger static RAM ICs have been developed since then. In most larger memories, each memory address location will store more than a single bit. For example, a 256 × 4 RAM would store 1024 (1 K) bits in the form of 256 independent words (or nibbles).

Dynamic RAM

The other basic type of RAM is dynamic RAM. In this type of memory, each bit is stored in a capacitor (or a semiconductor equivalent). A charged (full) capacitor represents a logic 1, while a discharged (empty) capacitor represents a logic 0.

A dynamic memory cell is much simpler than a static memory cell. Compare the dynamic memory cell in Fig. 18-4 with the static memory

371

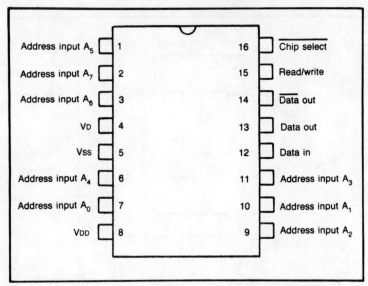

Fig. 18-3. One of the earliest static RAM ICs was the 256-bit MM1101.

cell that was shown in Fig. 18-2. This simpler circuitry means that a dynamic memory of a given storage capability will tend to be smaller and less expensive than a comparable static memory.

The MM5270 dynamic RAM IC, which is shown in Fig. 18-5 can store 4096 (4 K) independent bits of data. The 12 address lines allow for address values ranging from 0000 0000 0000 (decimal 0) to 1111 1111 1111 (deci-

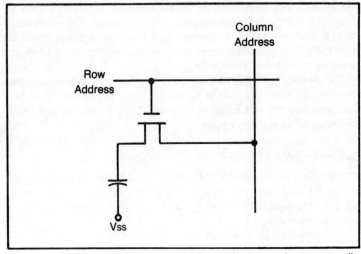

Fig. 18-4. A dynamic RAM cell is much simpler than a static memory cell.

372

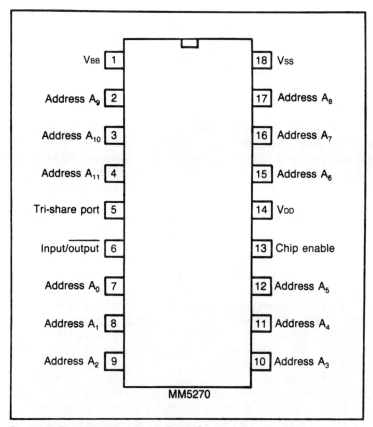

V_BB	1		18	V_SS
Address A_9	2		17	Address A_8
Address A_{10}	3		16	Address A_7
Address A_{11}	4		15	Address A_6
Tri-share port	5		14	V_DD
Input/output	6		13	Chip enable
Address A_0	7		12	Address A_5
Address A_1	8		11	Address A_4
Address A_2	9		10	Address A_3

MM5270

Fig. 18-5. The MM5270 dynamic RAM IC can store 4 K of data.

mal 4095). Dynamic RAM is certainly not without its disadvantages. The most important of these is the fact that no real-world capacitance can hold a charge indefinitely. Eventually, the charge will tend to leak off. Sooner or later, all of the 1s will become 0s. If left alone, a dynamic RAM will eventually clear itself.

Electronically reading the value stored in a dynamic memory cell tends to recharge the partially charged capacitors, refreshing the memory. Practical dynamic memory systems, therefore, require automatic refreshing circuits that will automatically read all of the memory locations at regular intervals, to prevent the charged storage capacitors from leaking off too much voltage.

A dynamic memory cell is simpler than a static memory cell, but dynamic memories require more complex supporting circuitry (to periodically refresh the capacitor charges). As with everything else in real life, a trade-off must be faced.

Recent improvements in solid-state technology have allowed for more static memory cells to be contained within a single IC chip, and at some-what lower manufacturing costs. This means that, at the moment, the balance of the scales tips somewhat in favor of static over dynamic memories in most general applications.

ROM

Data can be read out of, or written into a RAM. The user can store his own data — freely changing or updating it at any time. However, if the power to the system is cut off for any reason (often even for a fraction of a second), all data stored in RAM will be irretrievably lost.

In many applications there may be some data to be stored in memory that should be impossible for the user to inadvertently change or erase it. In computers, this includes the operating system for the computer system itself. Sometimes what we need is a memory that can be read from, but not written to. This type of memory is called a ROM, or read only memory.

All of the data in a ROM is permanently determined by the manufacturer when the chip is made. It can never be changed. Clearly, ROMs are only practical for applications where many identical units are required.

A typical application would be a ROM in a microcomputer that includes the instructions for translating BASIC commands into machine language. A ROM is highly desirable for this type of application. Few users want to have to load these instructions into the computer each time he turns it on. It should be permanent.

Since the data stored in a ROM is permanently hard-wired within the chip, each ROM memory cell can be made much simpler than either static or dynamic RAM cells. Several ROM cells are illustrated in Fig. 18-6. The data stored in each cell is determined by the presence (1) or absence (0) of a connecting diode at the appropriate address location. The user has no way of changing any of the data stored in a ROM. If even a single bit must be changed, the entire ROM chip must be replaced.

PROM

Since the data in a ROM must be determined at the time of manufacture, this type of memory is not very practical for applications where only a few copies are to be made. Unless you happen to need a few hundred identically programmed ROM chips, the cost per unit would be too high.

For applications in which permanent storage capabilities are required, but a manufacturer-programmed ROM would not be practical, a user programmable ROM has been developed. This type of memory is called *PROM*, or *programmable read only memory*. Each memory cell in a PROM is quite similar to the cells in a ROM. The PROM is manufactured with a connecting diode at each and every memory location. In other words, the memory contains all 1s.

The connecting diodes in a PROM are a special fused type. The user

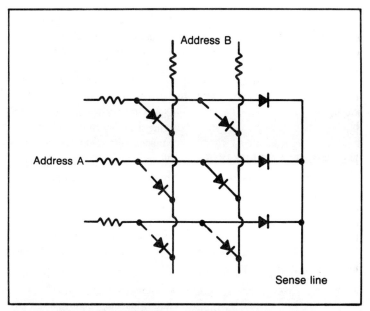

Fig. 18-6. The data in a ROM cell can never be altered.

can program the chip by applying a relatively large voltage to the unwanted diodes. The fuse element is broken by this procedure, so the cell's contents become a 0.

Once programmed, a PROM behaves exactly like a ROM. Data can be read from it, but the stored data cannot be erased or changed. Actually, more of the diodes could be blown (1s changed to 0s), but there is no way to replace a blown diode (change a 0 back to a 1). The voltage required to blow one of the fused diodes is higher than any signal which will normally be found in the computer system. A special PROM programming circuit must be used to blow the unwanted diodes. It's virtually impossible to accidentally blow a diode under normal operating conditions.

Once programmed, the data stored in a PROM cannot be changed. This means that if a mistake is made in programming, or even a small change must be made, the entire PROM must be discarded, and a new one must be programmed from scratch.

EPROM

A special type of PROM does allow the chip to be reused. All data can be erased, and new data can be written into the chip. Not surprisingly, this type of memory is called an *erasable programmable read only memory*, or *EPROM*.

An EPROM works in essentially the same way as a regular PROM, except for the fact that the entire chip can be cleared (all of the stored data is erased) by exposing the device to a strong ultraviolet light source. Note that there is no way to change just a few bits. It is an all or nothing proposition. Sunlight and other visible light sources contain some ultraviolet energy, so a programmed EPROM should be shielded from all light to avoid accidental erasure of the data.

Another type of reusable PROM is the *EEPROM*, or *electrically erasable programmable read only memory*. In this type of device, a special electrical signal is used to erase the entire chip, instead of exposure to ultraviolet light.

MEASURING MEMORY SIZE

Because binary numbers are used to define the memory location addresses, the number of cells in a digital memory system is always a power of 2, such as 256, 1024, or 4096. In large systems, memory size is usually defined as being so many *K*. *K* is normally used to indicate a factor of one thousand. However, 1000 is not a power of 2. The nearest power of 2 is 1024, so in memory systems, *K* actually represents a factor of 1024. Here are a few typical examples:

$$
\begin{aligned}
1 \text{ K} &= 1024 \\
4 \text{ K} &= 4096 \\
16 \text{ K} &= 16{,}384 \\
64 \text{ K} &= 65{,}536
\end{aligned}
$$

You do have to be careful about just what is being counted. For most memory ICs, 1 K indicates a storage capability of 1024 bits, while for most computer systems the quantity in question is the number of bytes (or 8-bit binary words) that can be stored. A 4 K memory in this case would hold 4096 sets of 8-bits each, or a total of 32,768 individual bits (0s and 1s). The usage of this terminology can cause some confusion if you're not careful.

APPLICATIONS

The most obvious and familiar application for memory devices is in a computer system, but this is most definitely not the only possible application. Memory devices can be used in many digital circuits that might require some degree of programmability, or a specific sequence of digital signals (bits). For example, a memory circuit might drive a series of logic gates to cause certain events at specific times. For this type of application, a clock/counter combination is used to step through the memory's address locations.

Index

Edited by Roland S. Phelps

Other Bestsellers From TAB

☐ THE LINEAR IC
HANDBOOK—Morley

Far more than a replacement for manufacturers' data books. *The Linear IC Handbook* covers linear IC offerings from all the major manufacturers—complete with specifications, data sheet parameters, and price information—along with technological background on linear ICs. It gives you instant access to data on how linear ICs are fabricated, how they work, what types are available, and techniques for designing them. 624 pp., 366 illus. 6" × 9".
Hard $49.50   Book No. 2672

☐ THE 555 IC PROJECT
BOOK—Traister

Now, with this exceptionally well-documented project guide, you can begin to use the 555 timer IC in a variety of useful projects that are both easy and fun to put together. From an IC metronome to a ten-second timer and even an electronic organ, this is an exciting collection of 33 different projects that even a novice at electronics can make easily, inexpensively, and in only a few hours time. Most important it provides the know-how to start you designing your own original 555 IC projects to suit your own special needs. 224 pp., 110 illus.
**Paper $11.95   Hard $18.95
Book No. 1996**

☐ THE ENCYCLOPEDIA OF
ELECTRONIC CIRCUITS—Graf

Here's the electronics hobbyist's and technician's dream treasury of analog and digital circuits—nearly 100 circuit categories . . . over 1,200 individual circuits designed for long-lasting applications potential. Adding even more to the value of this resource is the exhaustively thorough index which gives you instant access to exactly the circuits you need each and every time! 768 pp., 1,782 illus. 7" × 10".
Paper $29.95   Book No. 1938

☐ BASIC INTEGRATED
CIRCUITS—Marks

With the step-by-step guidance provided by electronics design expert Myles Marks, anyone with basic electronics know-how can design and build almost any type of modern IC device for almost any application imaginable! In fact, this is the ideal sourcebook for every hobbyist or experimenter who wants to expand his digital electronics knowledge and begin exploring the fascinating new possibilities offered by today's IC technology. 432 pp., 319 illus.
**Paper $16.95   Hard $26.95
Book No. 2609**

☐ 101 SOUND, LIGHT AND
POWER IC PROJECT—
Shoemaker

At last! Here's an IC project guide that doesn't stop with how and why ICs function . . . it goes one step further to give you hands-on experience in the interfacing of integrated circuits to solve real-world problems. Projects include sound control circuits such as alarms and intercoms; light control projects from photoflash slave to a monitor/alarm; power control units and much more! 384 pp., 135 illus.
**Paper $16.95   Hard $24.95
Book No. 2604**

☐ DESIGNING IC CIRCUITS . . .
WITH EXPERIMENTS—Horn

With this excellent sourcebook as your guide, you'll be able to get started in the designing of your own practical circuits using op amps, 555 timers, voltage regulators, linear ICs, digital ICs, and other commonly available IC devices. It's crammed with practical design and construction tips and hints that are guaranteed to save you time, effort, and frustration no matter what your IC application needs. 364 pp., 397 illus.
**Paper $16.95   Hard $24.95
Book No. 1925**

Other Bestsellers From TAB